Volkmar Stenzel | Nadine Rehfeld

Functional Coatings

Volkmar Stenzel, Nadine Rehfeld: Functional Coatings

ISBN: 978-3-86630-856-5

Cover: Fraunhofer-IFAM

Volkmar Stenzel, Nadine Rehfeld
Functional Coatings
Hanover: Vincentz Network, 2011
EUROPEAN COATINGS TECH FILES
ISBN 3-86630-856-6
ISBN 978-3-86630-856-5

Please ask for our book catalogue
Vincentz Network, Plathnerstr. 4c, 30175 Hanover, Germany
T (202) 684-6630, F (202) 380-9129
E-mail: books@american-coatings.com, www.american-coatings.com

Layout: Vincentz Network, Hanover, Germany

ISBN 3-86630-856-6
ISBN 978-3-86630-856-5

European Coatings Tech Files

Volkmar Stenzel | Nadine Rehfeld

Functional Coatings

Volkmar Stenzel, Nadine Rehfeld: Functional Coatings

ISBN: 978-3-86630-856-5

Foreword

Just about every article or object we encounter in daily life has a surface that has been treated in some way during its manufacturing process. The surface is what we first recognise when we see or touch an article or object. Surface treatment is responsible for the decoration, surface feel and protection of the surface, including corrosion protection. New developments (e.g. nanotechnology, encapsulation technologies, etc.) have opened up new opportunities for the integration of new surface functions, in addition to decoration and protection.

One of the best known functional surfaces is the surface of lotus leaves. These surfaces are extremely dirt-repellent due to their hydrophobic properties. A large amount of research is also being conducted on other, equally interesting surface properties, e.g. anti-fouling, drag-reducing, and self-healing, and is striving to transfer the underlying surface principles into concepts for coatings. Interdisciplinary work groups in research organisations and industrial companies are actively exploring the enormous potential of functional coatings.

This book summarises the vast number of new developments and the results of this work.

The present book gives users of paint technology an overview of the most promising developments and approaches. It also gives readers an idea of what is currently possible and what is likely to be possible in the near future. It highlights the current status of development or market introduction of the different technologies and novel surface functions that go beyond decoration, corrosion protection and surface protection.

We naturally had the difficult task of selecting information from a huge number of publications and we had to focus on the most important aspects for the users of paints and coatings, paint chemists, and paint manufacturers. We chose to focus on organic coatings (including organic-inorganic hybrid coatings) and have not addressed PVD/CVD methods, galvanic methods and other non paint/lacquer related topics.

Bremen, Germany in October 2010

Volkmar Stenzel & Nadine Rehfeld

Volkmar.Stenzel@ifam.fraunhofer.de   Nadine.Rehfeld@ifam.fraunhofer.de

Volkmar Stenzel, Nadine Rehfeld: Functional Coatings

ISBN: 978-3-86630-856-5

Contents

Volkmar Stenzel, Nadine Rehfeld: Functional Coatings

ISBN: 978-3-86630-856-5

1 Introduction

The term "functional coating" has been widely used in recent years and is a buzz-word, as were "nano" and "bio" previously. Accordingly, a very large number of scientific papers have been published that present old and new approaches to functional surfaces and describe a huge number of potential applications for functional coatings.

What, though, do we mean by "functional coating"? There is in fact no general definition for this term. According to DIN EN ISO 4618:2006 a coating is in general a "...continuous layer formed from a single or multiple application of a coating material to a substrate". Furthermore, a coating material is defined as a "...product in liquid, paste or powder form which, when applied to a substrate, forms a film possessing protective, decorative and/or other specific properties". Functional coatings have special properties and so give the surface new additional functions. This meaning of the term is reflected in the content of this book.

So how can we utilise the great wealth of information? We chose to focus on organic coatings (including organic-inorganic hybrid coatings), and present the most important aspects for the users of paints and coatings, paint chemists, and paint manufacturers. It also gives readers an idea of what is currently possible and what is likely to be possible in the near future. It highlights the current status of development or market introduction of the different technologies and novel surface functions that go beyond decoration, corrosion protection and surface protection.

1.1 Motivation

Surface technology is used in all goods-producing sectors of industry. The added value from surface technology is approximately 3 to 7 %. Surface treatment, coating, and finishing usually focus on necessary functions such as corrosion protection, decoration, design, and surface protection. Additional surface functions can create additional value for industrial products, but what are the most important functions and which are most desired?

In order to answer these questions, the German Society for Surface Treatment (Deutsche Gesellschaft für Oberflächenbehandlung e.V.) carried out a survey ("Forschungsagenda Oberfläche", DFO Service GmbH, Neuss, 2007) to identify the needs of German industry in the area of surface technologies. More than 300 technical experts from about 100 companies, 30 institutes, colleges etc. and several industrial associations participated in this survey. Despite the fact that the survey only covered Germany, we believe the results also apply to most other advanced industrial countries.

From the many (more than 100) technical topics and ideas that were signalled by the survey, 3 clusters were formed:

- knowledge-based quality improvement
- efficient processes
- multifunctional surfaces

Volkmar Stenzel, Nadine Rehfeld: Functional Coatings

ISBN: 978-3-86630-856-5

From all the topics that were indicated, nine so-called flagship topics were identified. These flagship topics represent the research areas which are thought to have the highest impact on the competitiveness of manufacturing industry. The flagship topics were defined taking particular account of economic, environmental, and social sustainability.

The flagship topics are as follows:

- **active layers** (e.g. photovoltaic technology, catalytic surfaces)
- **switching surfaces** (e.g. switching between hydrophobic/hydrophilic behaviour, colours, electrical conduction/non-conduction)
- **anti-fouling surfaces** (e.g. lotus effect, photocatalytic self-cleaning, non-toxic maritime anti-fouling)
- **self-repairing surfaces** (long-term surface protection, e.g. wind turbines, heavy-duty corrosion protection)
- precision manufacturing via model-based control and regulation
- digital factory
- **hybrid materials with complex morphology** (e.g. anti-reflective glass coatings)
- rapid testing for degradation and corrosion
- **brand protection**

The bold letters indicate topics that are of relevance for the development of functional surfaces.

Further important findings of the survey were:

- The added value of surface and coating technology in Germany amounts to about 20 billion Euros per annum.
- Higher added value due to the introduction of new surface functions can considerably increase competitiveness. It has been estimated that a 5 % increase in added value due to innovative surface technology compensates a 20 % lower cost of manufacturing in foreign countries.

These two findings are motivation enough to be engaged in the development, manufacturing and selling of products with functional surfaces.

1.2 Content and aim of this book

This book outlines recent developments in the field of functional coatings, with the focus on organic-based materials. The first part describes so-called toolbox methods which are currently available to developers. These cover a large number of methods, ranging from the manufacture of specific topographies on the microscale and nanoscale to the use of microcapsules and nanoparticles, and customised surface immobilisation of molecules. The functionalities themselves are at the fore in the second part of this book. These can, for example, be structure-based (e.g. drag-reducing) or chemical-based (e.g. self-healing) and can be produced via very different routes (e.g. anti-fouling, anti-icing).

It is important to stress that most of the examples described here are not yet ready for widespread industrial implementation. Indeed, the functionalities that are described are at different stages of development. Some merely represent promising first laboratory results (e.g. surface binding of anti-freeze proteins), others are at the production concept stage (e.g. paint structuring for drag-reducing surfaces), and some are already in use (e.g. self-healing car lacquer systems). The book gives an indication of the stage of current developments in the area of organic-based coatings and what future technical applications can be expected.

2 Tool-box

2.1 Paint surfaces with defined microtopography and nanotopography

The topography of a surface plays a major role in determining a variety of properties of that surface. Besides obvious properties such as the degree of gloss/mat, others such as wettability, drag, adhesion, and light reflection are highly influenced by the surface topography. This fact can be utilised in order to generate functional paint surfaces. Indeed, nature has utilised this for millions of years, as illustrated by the following examples.

2.1.1 Examples from nature

Nature displays a wealth of different surface topographies in plants and animals, giving rise to surfaces having specific functions (see also Chapter 3.6). The fact that nature displays a great variety of functions via the creation of microstructures and nanostructures on surfaces demonstrates that there are opportunities for developing surfaces with similar functions for technical applications. Examples of natural functional surfaces based on microstructures and nanostructures are described in the following.

Lotus leaf

The self-cleaning effect of the lotus leaf [1] is certainly the best known example of a natural, functional surface. The same effect also occurs in other plants, for example lady's mantle and kohlrabi (albeit "kohlrabi effect" would sound far less interesting than lotus effect). The effect is based on chemical hydrophobicisation due to secreted waxes coupled with a microstructure and nanostructure (for mechanism see Chapter 2.1.2).

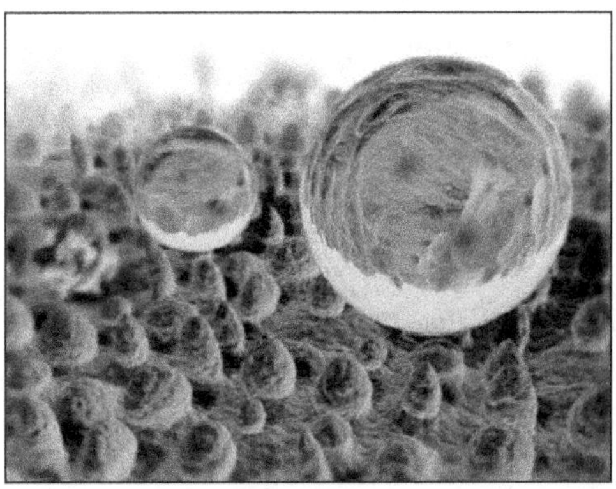

Figure 2.1: Surface of a lotus leaf (computer animation) source: S. Sepeur[2]

Skin of fast-swimming sharks

Decades ago it was observed that the scales of fast-swimming species of sharks have parallel riblets in the direction of flow around of the body. It is now known that this microstructure reduces the drag of the water by a few percentage points and so helps the shark save energy for motion (see Chapter 3.4).

Volkmar Stenzel, Nadine Rehfeld: Functional Coatings

ISBN: 978-3-86630-856-5

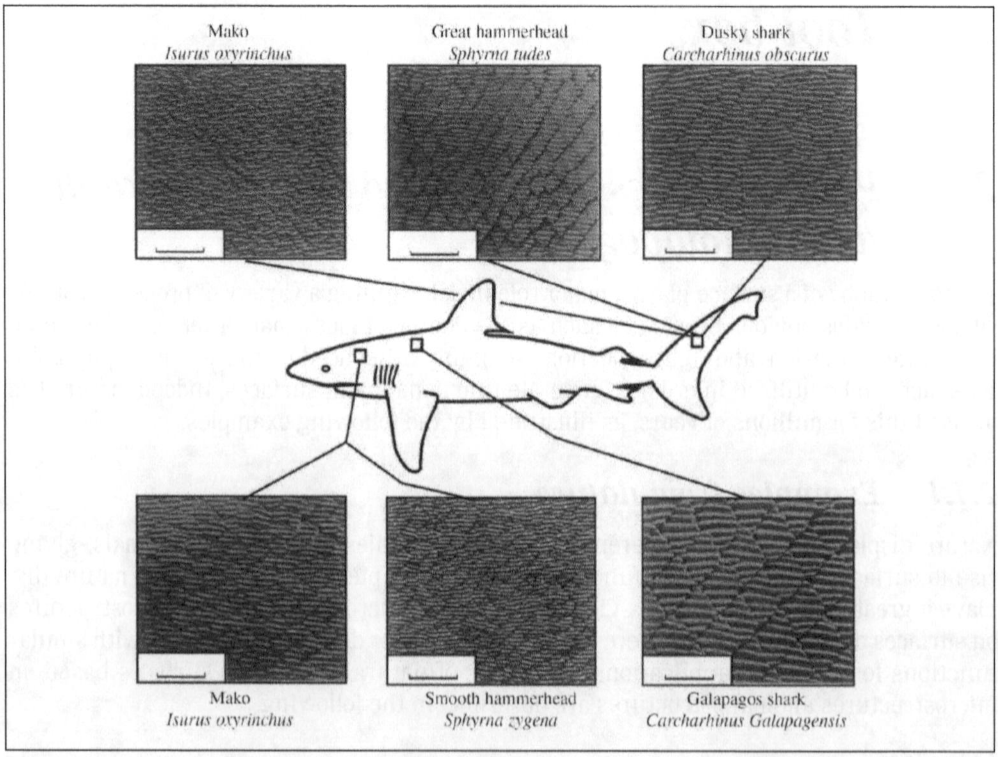

Figure 2.2: Surface structures on fast-swimming sharks [3] *source: Wolfram Hage, DLR*

2.1.2 *Influence of surface topography on wetting*

The wetting properties of a surface are very important for many surface functions. For example, the wetting properties very much determine the soiling, easy-to-clean properties, ice formation and adhesion, ability to be coated, and drag (in water). The surface topography at the microlevel and nanolevel in turn has a big influence on the wetting. For this reason, it is worth exploring the relationship between topography and wettability at this point in greater detail.

2.1.2.1 *Ideal surfaces*

The ability of a liquid to wet a smooth surface can be determined, for example, by measuring the contact angle. This is done by placing a drop of liquid on the surface and measuring the contact angle at the solid-liquid-air (or vapour) phase boundary using a special microscope (see Figure 2.3).

The contact angle of a liquid on a surface depends on the relationship between the surface energies (= surface tensions) of the relevant phases at the drop boundary (solid, liquid, gas). The relationship for an ideal smooth and homogenous surface is described by Young's equation [4]:

Equation 2.1 $$\cos\theta = \frac{\gamma_s - \gamma_{sl}}{\gamma_l}$$

where γ_s is the free surface energy of the solid, γ_l is the surface tension of the liquid, and γ_{sl} is the interfacial energy between the solid and liquid.

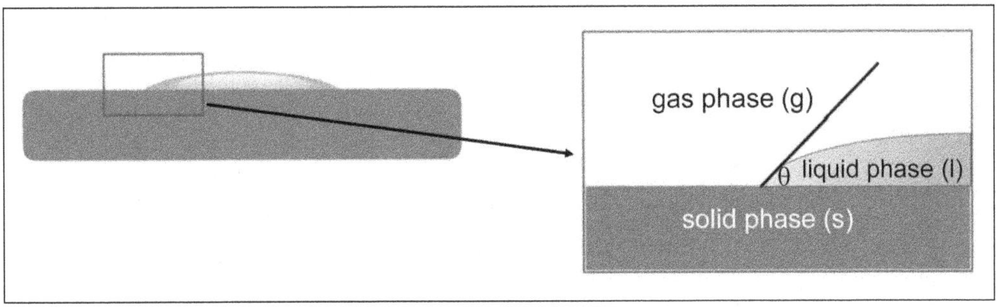

Figure 2.3 Definition of the contact angle θ

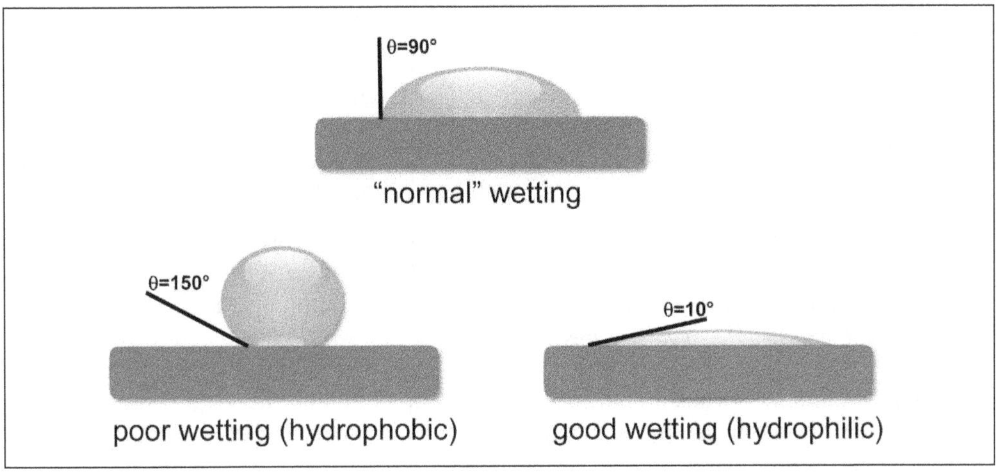

Figure 2.4: Different wetting behaviour of surfaces

If the contact angle θ is very small (< 10°), then the surface can be very effectively wetted (Figure 2.4). If the contact angle is close to 0, there is said to be complete wetting. If water is the wetting liquid, there is talk of ultrahydrophilicity. This occurs when the surface energy of the solid is equal to or larger than the surface tension of water. Under conditions when condensation can arise, no water droplets can form on such surfaces and they act as anti-fog surfaces.

Conversely, surfaces having a water contact angle of 140° or higher are termed superhydrophobic, namely extremely water-repelling (Figure 2.4). The surface energy of such surfaces is very much smaller than the surface tension of water. Water virtually no longer wets the surface at all. For water contact angles upwards of 160° the term lotus effect is often used. Such surfaces are also called easy-to-clean surfaces as they allow good run-off of water and often show reduced soiling.

Details and examples of such surfaces can be found in the nanotechnology book by Stefan Sepeur, which has also been published in this series [2].

2.1.2.2 Non-ideal surfaces

The aforementioned simple relationships only apply for ideal surfaces, namely surfaces that are ideally smooth and chemically homogenous. In the real world we generally deal with surfaces that do not meet these criteria. This becomes clear if we measure the contact angle in a dynamic way rather than statically. This can be done by measuring the contact angle as a function of advancing drop diameter (namely addition of liquid) and receding drop diameter (namely removal of liquid), see Figure 2.5.

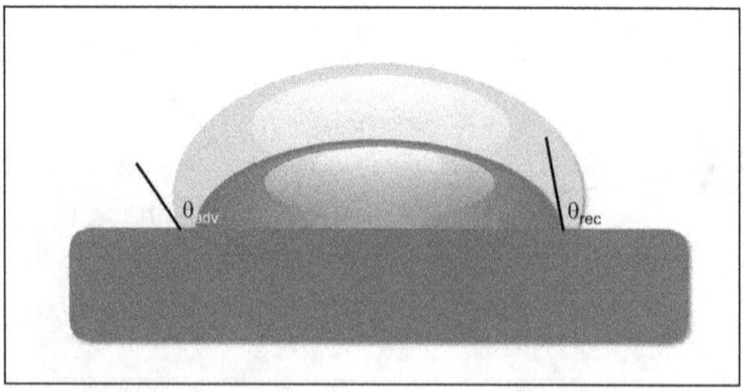

Figure 2.5: Advancing and receding contact angle; contact angle hysteresis

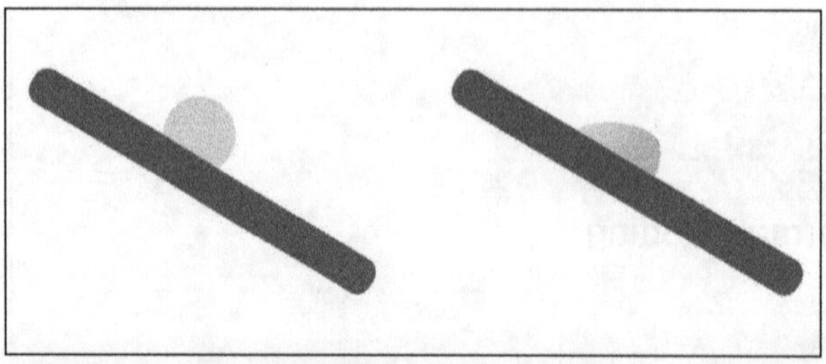

Figure 2.6: Drop on inclined surface; left: low contact angle hysteresis, right: high contact angle hysteresis

The contact angle for an advancing drop diameter is generally greater than for a receding drop diameter. This difference is called contact angle hysteresis and usually amounts to between 5 and 20°, but can be considerably higher [5].

Contact angle hysteresis is particularly evident when drops run off an inclined surface, with on the one side an advancing contact angle and on the other side a receding contact angle (Figure 2.6).

The reasons for the differences in the contact angle with advancing and receding contact angle include [6]:

- Topographic surface roughness. When the drop meets a bump on the surface, it jumps into a position having the same contact angle as previously attained. At this point the spreading out of the drop is hindered until it is large enough to overcome the bump. The same effect causes the reduction in size of the drop, and the result is the described hysteresis.
- Chemical heterogeneity of the surface. When the drop spreads out on such a surface, the three-phase boundary line is kept in place (pinned) by relatively liquid-repelling (lyophobic) regions, and when the drop recedes the boundary line is kept in place by liquid-attracting (lyophilic) regions. The result is again hysteresis.
- On non-rigid surfaces (e.g. polymers) the forces that act on the three-phase boundary line can be so large that the surface deforms, and this can also lead to contact angle hysteresis.

In reality these effects often occur in combination, and this can result in considerable intensification of the effect. The roughness has a particularly large effect on contact angle hysteresis, and for this reason this is considered in detail below.

In the past, two essentially empirical laws were put forward for describing the wetting behaviour of rough surfaces. These also represent the two cases that occur in nature [7] (see Figure 2.7).

Wenzel regime

Wenzel described an average contact angle on a rough, chemically homogenous surface [8]. He defined the contact angle on a rough surface θ* as a function of the Young contact angle θ (see above):

Equation 2.2
$$\cos\theta^* = r'\cos\theta = r'\,\frac{\gamma_s - \gamma_{sl}}{\gamma_l}$$

Wenzel introduced the roughness coefficient (r') which represents the relationship between the actual surface and the geometric projected surface. The roughness coefficient (r') is

Figure 2.7: Schematic representation of the two different cases for wetting rough surfaces, (a) the Wenzel regime and (b) the Cassie-Baxter regime source: Ulrike Mock [6]

always larger than 1. Hence in this wetting regime, when considering water as the wetting liquid, both the hydrophilicity and the hydrophobicity are enhanced by the roughness.

The other case is the Cassie-Baxter regime (see Figure 2.7) which has originally been described for smooth, chemically heterogeneous surfaces [9], but which also applies to rough surfaces where air is trapped under the wetting liquid in indentations. For this type of surface (comprising a solid and air), the contact angle θ^* is the average contact angle of the drop on air (assumed to be 180°) and on the (ideally smooth) solid (θ). If a surface has a wetted fraction ϕ_s (that is the fraction of the surface on which the liquid is in direct contact with the solid), the following equation applies:

Equation 2.3 $$\cos\theta^* = \Phi_s \cos\theta + (1-\Phi_s)\cos 180° = \Phi_s(\cos\theta + 1) - 1$$

The two wetting regimes have very different behaviour, in particular regarding contact angle hysteresis. When a Wenzel regime prevails, the drops have relatively large contact angle hysteresis and a larger run-off angle (the angle relative to the horizontal above which the drop runs off). In contrast, when a Cassie-Baxter regime prevails, the hysteresis and also the run-off angle are small.

The surface chemistry and the roughness largely determine which of the two regimes prevails. The typical behaviour of a given surface with low surface energy is as follows: First of all, the wetting with water follows the Wenzel regime. With increasing roughness (or roughness factor), the contact angle and hysteresis increase. Once a certain roughness value is exceeded, the contact angle increases further but the hysteresis decreases considerably. At this roughness value above which the hysteresis decreases, the system changes over from the Wenzel regime to the Cassie-Baxter regime, namely the fraction of air at the boundary increases considerably.

In the region of moderate hydrophobicity (Young contact angle up to 120°), the Cassie-Baxter regime is metastable. External pressure (in the simplest case, touching the water drop) can lead to the system switching irreversibly into the Wenzel regime. That means that the liquid now follows the surface contour and there is no longer any air between the water and the surface. The contact angles of both regimes are similar in this region, but the hysteresis in the Wenzel regime is 10 to 20 times higher. In the region of the Wenzel regime, the drops hence adhere much more strongly to the base surface.

Due to this behaviour, it often makes more sense to measure the run-off angle than the contact angle in order to characterise the wetting behaviour. In particular for surfaces whose functions are based on their wetting behaviour (e.g. easy-to-clean layers or special anti-icing layers), it is vital to consider both the contact angle and run-off angle when characterising the surface properties.

Customising the hydrophobicity or hydrophilicity of a surface with a given chemistry by introducing a special microstructure or nanostructure is essential for manufacturing surfaces with particular functional properties. The highest known water contact angle of a (smooth) material is 120° [10]. This involves hexagonal close packing of CF_3- groups. If one wants to achieve higher contact angles, such a surface must be combined with a suitable topography. Using this principle (combining hydrophobic surface chemistry with suitable microtopography and nanotopography – the lotus principle), superhydrophobic surfaces with contact angles > 150° can be produced. Here, microstructures are often combined with nanostructures, as is the case in nature with lotus leaves.

The ability of a suitable topography to make hydrophilic surfaces more hydrophilic and also hydrophobic surfaces more hydrophobic is also utilised for switchable surfaces. For example, the switching amplitude of the water contact angle for a surface modified with polymer brushes (see Chapter 2.5) was increased from 20° for a smooth surface to 150° when the surface had a suitable microstructure and nanostructure [11].

Microstructures or nanostructures can be either stochastic (namely random and without a recognisable uniform pattern) or deterministic. Deterministic here means that the surface has a very special, uniform pattern that is responsible for the specific function of the surface.

2.1.3 Creation of stochastic surface structures in paint films

The creation of surfaces with rough topographies is a classical technique for making mat paints, amongst other things. In the simplest case the paint contains solid, particulate matting agents. After the drying process, which involves loss of solvent and curing shrinkage of the binder and is accompanied by a reduction in volume of the paint film, the result is microscopic spots sticking out of the paint film. Figure 2.8 shows the principle.

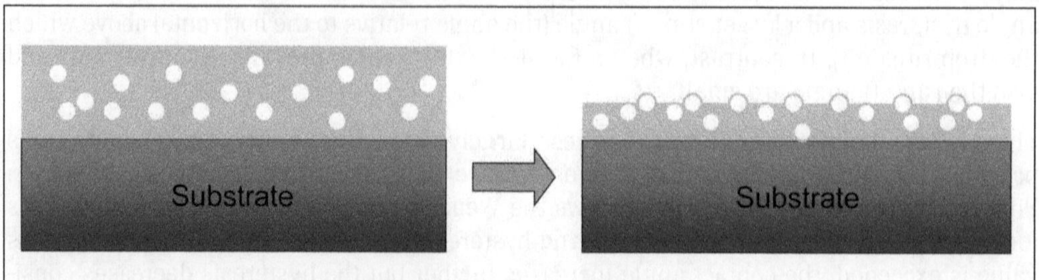

Figure 2.8: Left: wet paint film, freshly applied; right: paint film after drying

Figure 2.9: Stochastic surface topography created by microparticles source: Fraunhofer IFAM

The size of the particles determines the scale of the surface topography. Depending on the particle size, surface structures can be produced on a microscale or nanoscale. Figure 2.9 shows a surface produced in this way.

Using the same principle it is also possible to manufacture hierarchical surface topographies, for example those that combine a microstructure with a nanostructure. The relevant particle sizes are combined in the paint film for this purpose (Figure 2.10).

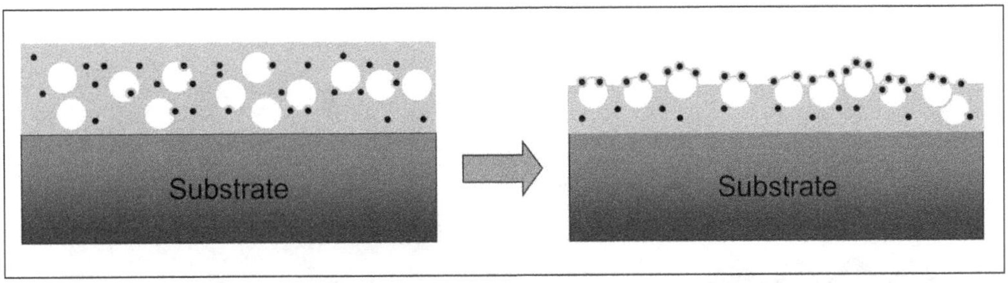

Figure 2.10: Left: wet paint film, freshly applied; right: paint film after drying

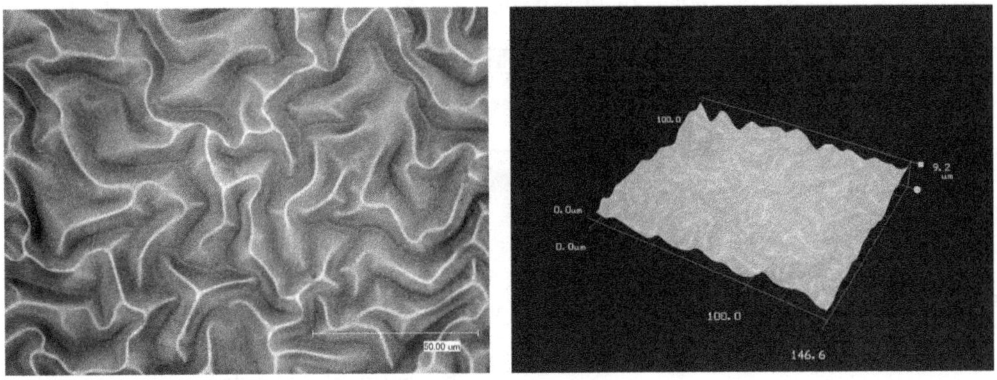

Figure 2.11: Surface topography created using a two-stage UV-curing method *source: Fraunhofer IFAM*

By suitable choice of the particles, the coating matrix, and the volume concentration of the pigment, it is possible to produce paint surfaces having a water contact angle of >140°. So-called lotus effect coatings or paints are usually produced in this way.

The customisation of the topography using microfillers and nanofillers is generally successful for paints that shrink significantly in volume on drying. For paints where this is not the case, other methods are available.

UV-cured coatings generally do not shrink in volume due to solvent loss on curing. There is solely curing shrinkage due to the crosslinking reaction, which can amount to up to 10 %. In order to adjust the surface structuring in such systems, the curing can be undertaken in two stages with different wavelengths. The first curing stage (small wavelength, low penetration depth) leads to crosslinking of the uppermost microns of the paint film. Due to the curing shrinkage in the uppermost layer, which almost floats on the uncured liquid coating below, small folds form and these create the desired surface topography. The subsequent curing at higher wavelength leads to full curing of the coating film. By varying the process parameters (layer thickness, wavelengths, curing times), the topography can be customised over broad limits. Figure 2.11 shows a surface produced using this principle.

This method is much used in the graphics industry for producing mat printing inks.

2.1.4 Creation of deterministic surface structures

The before-described methods are very suitable for producing stochastic surface structures, namely random surface structures. In order to introduce specific roughness, for example to customise the wetting properties, these methods are excellent. Other methods must, however, be used when defined, regular structures are required for specific surface functions.

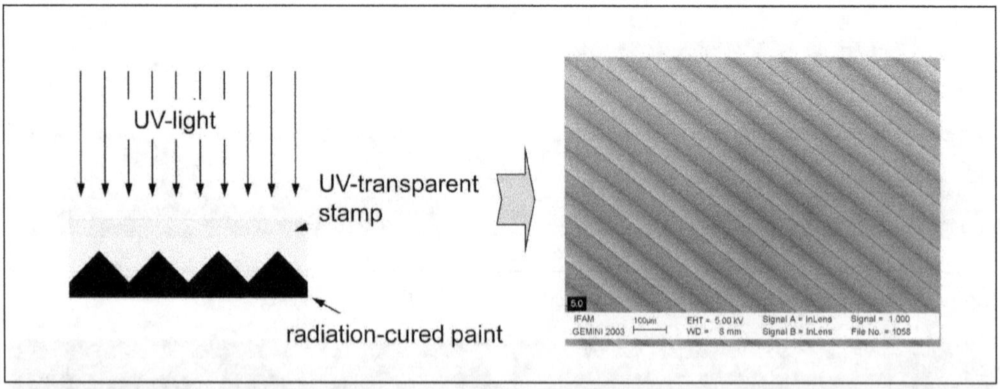

Figure 2.12: Left: principle of simultaneous embossing and curing; right: microstructure in a UV-cured paint *source: Fraunhofer IFAM*

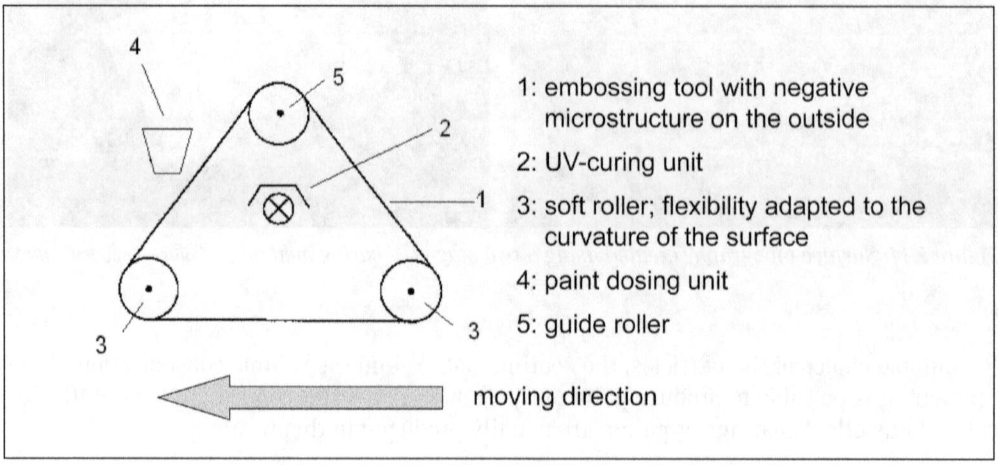

Figure 2.13: Principle of the roller-tool for paint application, embossing, and curing [12]

In the simplest case one can imagine the paint film being embossed with a suitable tool during the drying process. This assumes that during the drying process the paint film has a time window when the paint is free of solvent but is not so fully cured that cannot be embossed, and that after removing the tool the structure remains, is not tacky and does not stick to the tool.

A viable time window is generally not available for standard paints. For this reason, an alternative is to carry out the curing and embossing of the surface in a single step, so that the paint loses its tackiness during the curing and there is no reflow of the structures after removal of the tool. This can be effectively achieved using radiation curing paints which can be formulated without solvents (with the exception of reactive thinners). The tool here is preferably made of a material that is transparent to the radiation used for the curing and a material that adheres as little as possible to the cured paint film. The principle is depicted in Figure 2.12.

A stamp method, as described above, can naturally be used for small surfaces that can be stamped by hand or by machine. For larger surfaces, for example facade elements, aircraft components, or ship hulls, a continuous process must be employed. Figure 2.13 shows the principle of a continuous paint application and curing process which is currently being developed for industrial use.

A paint film consisting of an (at least partially) UV-curing polymer system is applied using a wide slit nozzle (4) to a moving UV-transparent belt (1) bearing the negative of the desired structure. When the tool is rolled across a surface, the paint film ends up below a soft roller (3) and is transferred to the surface to be coated. When the paint then comes into the field of the UV lamp (2), it is crosslinked until at least the paint film does not run anymore and is touch-dry. Decisive is that the paint loses its tackiness and does not adhere to belt. The tool therefore leaves behind a microstructured or nanostructured paint film on any desired surface. The application tool can, for example, be controlled using a robot (see Figure 2.14).

Figure 2.14: Application of microstructured or nanostructured paint films using a robot
source: Fraunhofer IFAM

An advantage of this method, compared to the application of films, is that single or multiple curved surfaces can also be coated, because the embossing film and the soft rollers can be customised in a very versatile way (see Figure 2.15).

Virtually any desired nanotopography or microtopography can be applied to large surfaces using this method. With careful selection of the material used for the embossing tool, undercuts can even be reproduced to a certain degree. Figure 2.16 shows two examples of paint surfaces produced with radiation curing paint systems using the aforementioned principle.

Other examples of paint surfaces produced using this method are depicted in Figure 2.17.

The method described here can obviously only be used for reproducing microstructures and nanostructures. The manufacture of the relevant master molds must be undertaken with other, sometimes complex, methods including micro-machining [13], lithographic methods [14] and laser technologies [15].

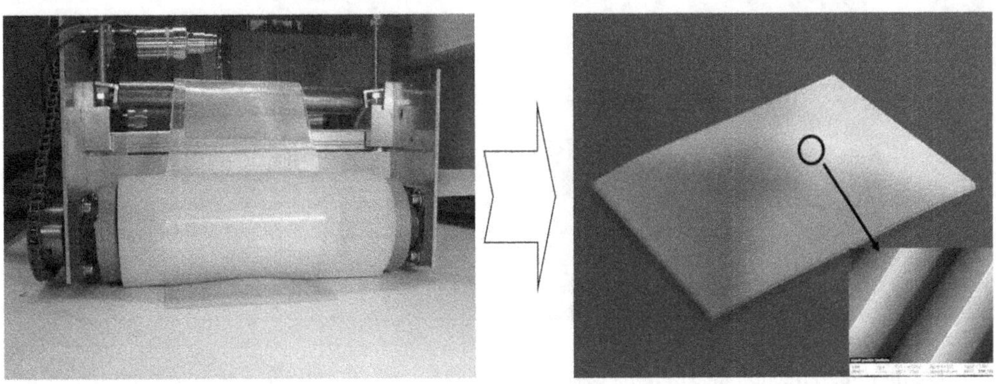

Figure 2.15: Application of a structured paint film on double-curved surfaces *source: Fraunhofer IFAM*

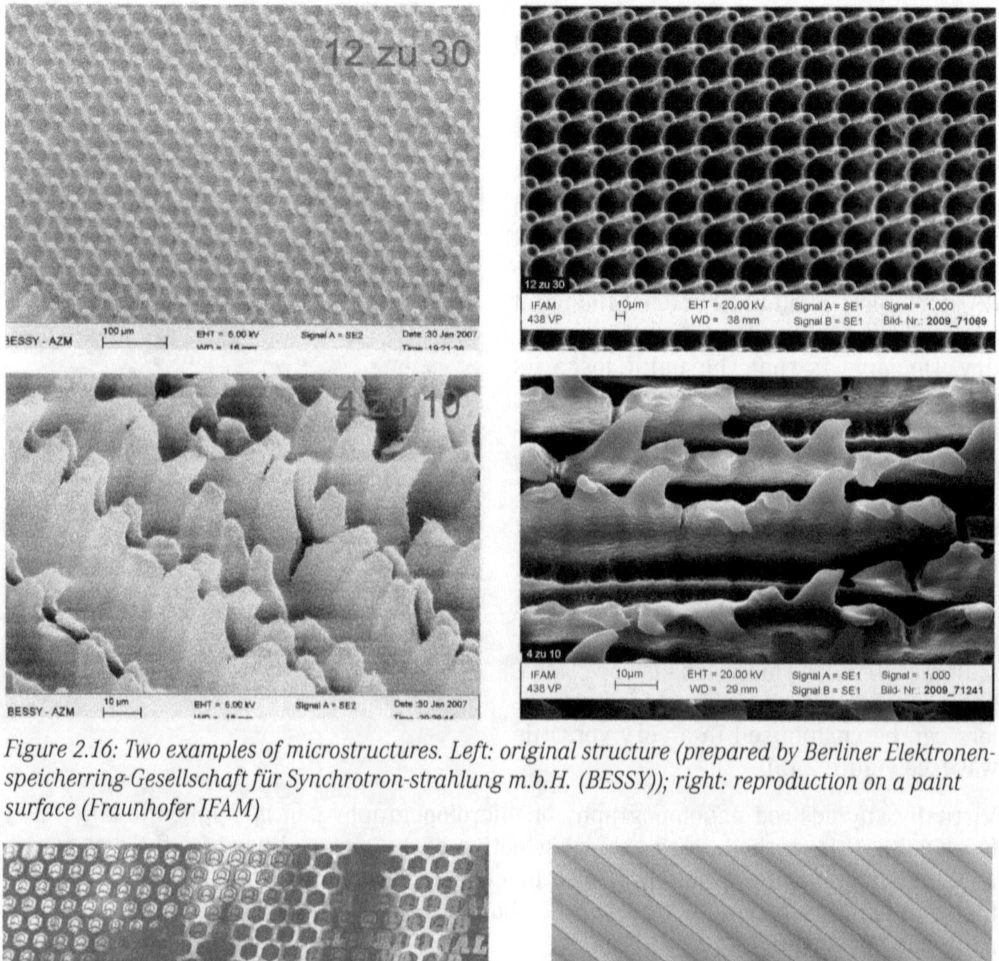

Figure 2.16: Two examples of microstructures. Left: original structure (prepared by Berliner Elektronen-speicherring-Gesellschaft für Synchrotron-strahlung m.b.H. (BESSY)); right: reproduction on a paint surface (Fraunhofer IFAM)

Figure 2.17: Left: hologram in a clearcoat (master-mold courtesy of topac GmbH); top right: riblet structure for drag reduction (see Chapter 3.4), bottom right: anti-reflective nanostructure (master-mold courtesy of Holotools GmbH)

2.1.5 Literature

[1] *Barthlott, W.;* Scanning electron microscopy of the epidermal surface in plants, Systematics Association's Special, 41, 1990, S. 69-94

[2] *Sepeur, S.*; Nanotechnology, Technical Basics and Applications, Vincentz Network, 2008

[3] *Hage, W.*; Zur Widerstandsverminderung von dreidimensionalen Riblet-Strukturen und anderen Oberflächen Dissertation Berlin 2004

[4] *Young, T.*; Philosophical Transactions of the Royal Society of London, 95, 1805, 65-87

[5] *Butt, H.-J.; Graf, K.; Kappl, M.*; Physics and Chemistry of Interfaces, Wiley-VCH Verlag GmbH & Co. KGaA, Weinheim, 2003

[6] *Mock, U.*; Dissertation „Über das Benetzungsverhalten polymermodifizierter Grenzflächen", Freiburg, 2004

[7] *McHale, G.; Cassie and Wenzel:* Were they really so wrong?, Langmuir, 23, 2007, 8200-8205

[8] *Wenzel, R. N.*; Journal of Physical and Colloid Chemistry, 53, 1949, 1466-1467

[9] *Cassie, A. B. D.; Baxter, S.*; Transactions of the Faraday Society, 40, 1944, 546-551

[10] Nishino, T.; Meguro, M.; Nakamae, K.; Matsushita, M.; Ueda, Y.; The lowest surface free energy based on $-CF_3$ alignment, Langmuir, 15, 1999, 2551-2558

[11] *Uhlmann, P.; Houbenov, N.; Ionov, L.; Motornov, M.; Minko, S.; Stamm, M.*; Oberflächen passen sich an - bürstenartige Polymermoleküle an Oberflächen mit schaltbaren Eigenschaften, Wissenschaftliche Zeitschrift der Technischen Universität Dresden, 56, 2007, 47-52

[12] Patents DE_10346124_B4 and DE_102006004644_B4

[13] *Brinksmeier, E.; Gläbe, R.; Riemer, O.; Twardy, S.*; Potentials of precision machining processes for the manufacture of micro forming molds, Mikrosystem Technologies, 14 (12), 2008, 1983-1987

[14] www.holotools.de

[15] *Römer, G. R. B. E.; Huis in't Veld, A. J., Meijer, J.; Groenendijk, M. N. W.*; On the formation of laser induced self-organizing nanostructures, Manufacturing Technology, 58, 2009, 201-204

2.2 Microcapsules

2.2.1 Introduction

Industrial use of microcapsules has grown very fast over the last 30 years. One of the first industrial applications for microcapsules was carbonless copy paper. In 1974 approximately 500,000 tons of this paper was produced, corresponding to 50,000 tons of microcapsules [1].

Microcapsules are used in the food industry (e.g. microcapsules containing flavours in chewing gum), in the detergent industry (e.g. microencapsulated perfume in fabric-softeners which is released weeks after the washing process when the clothes are touched and worn), and in the pharmaceutical industry. The reasons for using microcapsules in these fields are:

- Protection of sensitive materials against environmental attack, enhancement of shelf-life;
- Controlled, triggered, or delayed release of active agents;
- Masking of smell or taste;
- Improved processability (flow properties, handling of toxic materials, and handling of liquids as solids).

In the above-mentioned industries, microcapsules are very common raw materials. Using microcapsules must, however, not increase product prices for the consumer too much. In the food industry, for example, a maximum cost for the microencapsulation can be roughly estimated at €0.1/kg [2]. This figure shows that microencapsulated raw materials for potential use in functional coating materials must be within an acceptable price range.

2.2.2 Microencapsulation techniques

The technique of microencapsulation aims at enveloping liquids (in special cases gases and solids too) in a finely dispersed form. The particle diameter can range between 1 and 5000 µm. Typical capsule diameters for different applications are as follows [3]:

- 1 to 10 µm for carbonless copy paper;
- 30 to 50 µm for pesticides;
- 10 to 50 µm for perfumes.

The thickness of the shell typically ranges from 50 nm to several µm. The shell can be formed from synthetic polymers and from biologic materials as well. The external shape of the capsules is mostly spherical. In special cases, grape-like clusters or irregular structures appear. The nature of the wall material and wall thickness determines the ability of the microcapsule to protect the core material or to release it.

The capsule wall separates the content from the external surroundings, and to release it the shell must be opened or must be permeable. The opening may occur via mechanical stress from the outside (shearing, crushing) or via effects from the inside (heating above the boiling point, melting, explosion). Some examples are given below.

For the microencapsulation of agents, numerous preparation technologies are available. In general, microencapsulation techniques can be divided into chemical and physical methods. The Table 2.1 gives an overview of current technologies that are widely used on a laboratory scale and for the industrial production of microcapsules [4].

For application of microcapsules in paints and coatings, the particle size should not exceed approximately 50 µm. For many applications (primers, spray-coatings, etc.) the capsule size

Table 2.1: Overview of microencapsulation technologies

Chemical processes	Physico-chemical processes	Physico-mechanical methods
Interfacial polycondensation Emulsion polymerisation	Coacervation Layer-by-layer assembly Sol-gel encapsulation Supercritical CO_2-assisted microencapsulation	Spray drying Dipping or centrifuging techniques Co-extrusion Fluidized bed technology

should be between 1 and 20 µm. Some applications require even smaller capsules. In the following chapters several examples of encapsulation techniques that can produce capsules in the relevant size range are explained in more detail.

2.2.2.1 Interfacial polycondensation, polyaddition or radical polymerisation

A very versatile method for encapsulation of active agents that form a separate, water-immiscible phase in water is interfacial polycondensation or polyaddition. The basis for the microencapsulation process is an emulsion of the oil-in-water (o/w) type. During the microencapsulation process a wall forms around the oil-phase droplets from appropriate monomers via a crosslinking process. The process is outlined in Figure 2.18.

The microencapsulation process comprises the following steps:

Choice of the organic (dispersed) phase

The organic phase contains the active agent for the function that is going to be introduced to the coating material via the microcapsules. Examples of active agents are given below (Chapter 2.2.4). The organic phase must have very low solubility in water and should have a viscosity that allows the preparation of an emulsion in water at a suitable temperature.

Preparation of the water (continuous) phase

The continuous phase contains water-soluble components (additives) such as:

- pH-regulators
- Defoamers
- Agents that influence the wall stability or transparency
- Components of the capsule wall

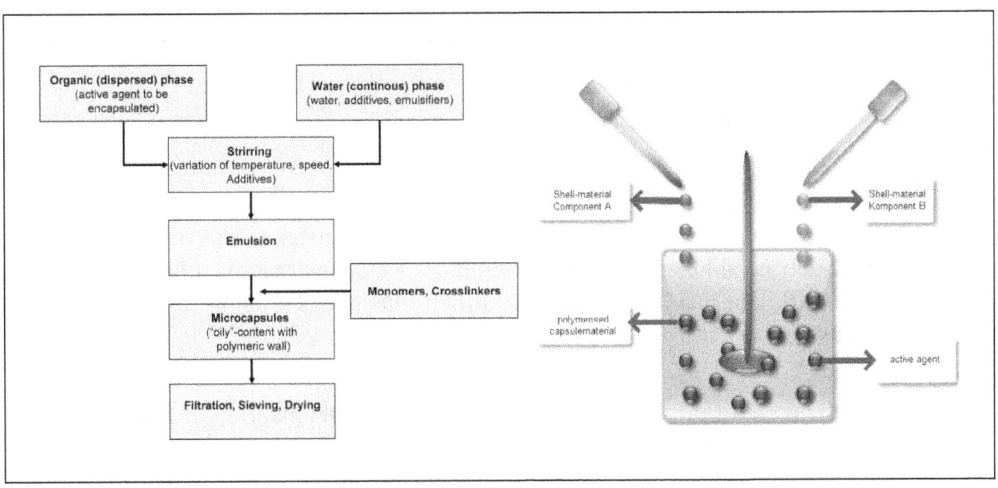

Figure 2.18: General principle of microencapsulation by interfacial reaction

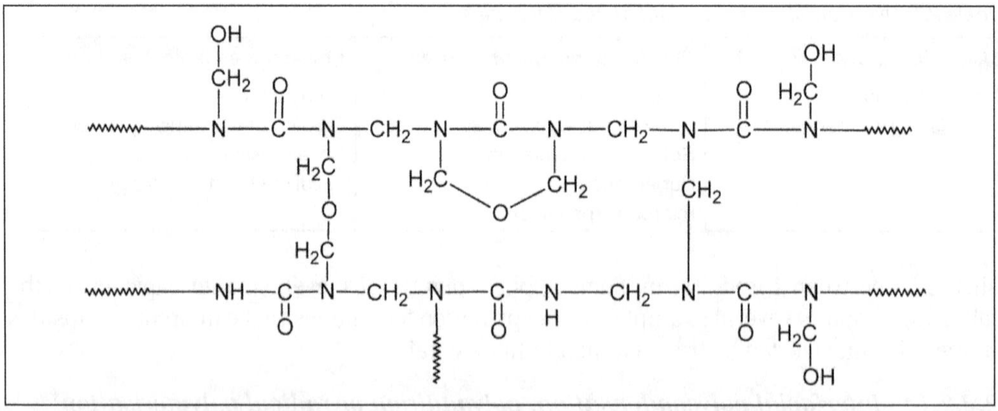

Figure 2.19: Poly(urea-formaldehyde) capsule material

Preparation of the emulsion

The emulsion is the basis for the microencapsulation process. The droplet size that is achieved during the emulsification determines the size of the microcapsules. The droplet size mainly depends on the following parameters:

- Stirring conditions (speed, type of mixer, mixing geometry)
- Type of emulsifier
- Amount of emulsifier

If stirred by conventional means, the size of the microcapsules ranges from approximately 2 µm up to several hundred microns. By employing a sonication device during the emulsification process, particle diameters down to 220 nm can be achieved [5]. Another method for achieving fine particles with a narrow size distribution is the Shirazu porous glass (SPG) emulsification technique [6].

Wall-building

After adding the monomers and adjusting the appropriate reaction conditions (e.g. pH, temperature) the wall-building process starts. The wall-building reaction can be a polycondensation, polyaddition or radical-polymerisation process. The monomers, crosslinking agents, catalysts, etc. must be soluble in at least one of phases. The polymeric wall-material has to be insoluble in both phases. The achieved wall thickness depends on the concentration of monomers and the reaction conditions. Typical values for the wall thickness of microcapsules produced by interfacial reaction are between 50 and 150 nm.

Several quite different chemical reactions have been used to create microcapsules via an interfacial polymerisation process, e.g.:

- Poly(urea-formaldehyde) (PUF) [7],[8]
  The wall-building reaction involves polycondensation of urea and formaldehyde. This reaction is widely used for encapsulation of resins or monomers for self-healing systems (see Chapter 3.3). The reaction takes place under moderate temperatures (<80 °C) and leads to a capsule-polymer as shown in Figure 2.19.

- Poly(melamine-formaldehyde) (PMF) [9]
  In this case the wall-building reaction involves polycondensation of melamine and formaldehyde. The process and the chemical nature of the polymer are quite similar to PUF. Capsule walls made of PMF have different mechanical properties in terms of their strength and their interactions with polymeric matrices of coating systems.

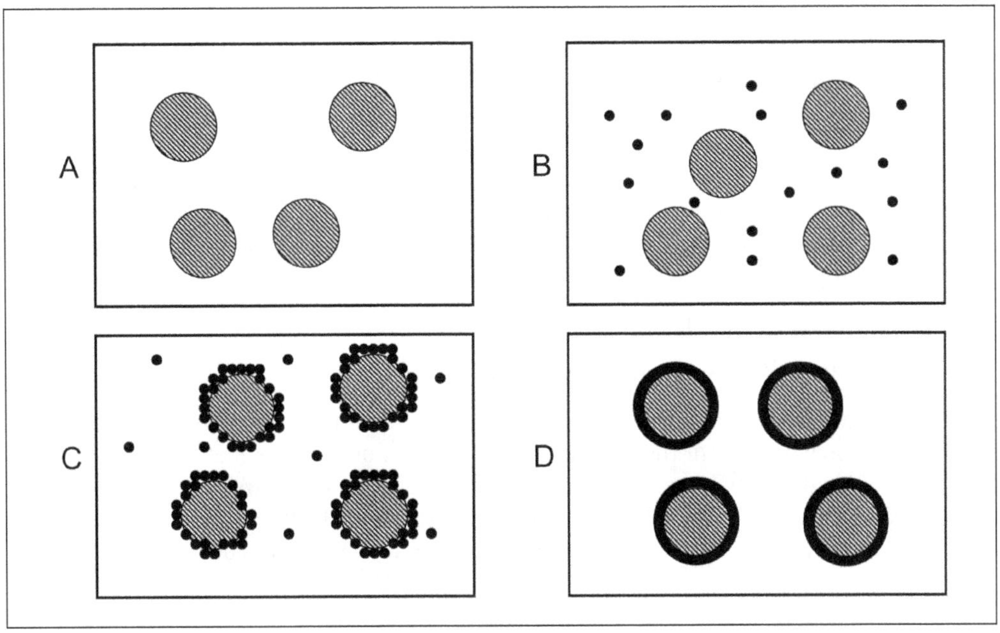

Figure 2.20: Principle of microencapsulation by coacervation. A: Dispersed core material in a continuous phase which contains the wall materials (e.g. gelatin/gum arabic) in solution; B: Onset of coacervation by precipitation of fine-particle microcoacervate from solution; C: Precipitation of microcoacervate on core material droplets; D: Coalescence of the microcoacervate on the wall material phase [1]

- Polymerisation products of epoxy resins and carboxylic acids [10]
  A versatile process for the formation of capsule walls is the reaction of an epoxy resin with bi- and tri-functional carboxylic acids. The choice of different acids allows optimisation of the properties of the polymer for the desired application. A microencapsulation process using this reaction has been reported recently, where a diglycidyl ether of bisphenol A was crosslinked with decanedioic acid and 1,3,5-benzenetricarboxylic acid. This process led to stable microcapsules with diameters ranging from 10 to 400 μm depending on the reaction conditions.

- Polyurea [3], [11]
  The reaction of difunctional isocyanates with bi- or trifunctional amines or with hydrolysed isocyanates leads to the formation of polyurea. In this case, microcapsules are formed in an emulsion where the organic (dispersed) phase contains the isocyanate, e.g. hexamethylene-1,6-diisocyanate (HMDI) or toluene-2,4-diisocyanate (TDI); and the continuous phase contains the amine component, e.g. hexamethylene-1,6-diamine (HMDA) or diethylenetriamine (DETA). The reaction occurs rapidly at ambient temperature and leads to microcapsules in the range of 0.5 to 50 μm.

A microencapsulation process for water-soluble active agents is also possible. In this case an inverse process can be used, in which the dispersed phase is water-soluble and the continuous phase is organic. In this case the microencapsulation takes place in a w/o-type emulsion. One example is given below:

- Crosslinking of hydrophilic polymers e.g. poly(vinyl alcohol) [12]
  The crosslinking of hydrophilic polymers, for example the reaction of poly(vinyl alcohol) with glutaraldehyde, can be used to produce semipermeable microcapsules that can be used for controlled release of active agents. The capsule size for this process ranges from 2 to 55 μm.

The microencapsulation via interfacial polymerisation is a very versatile method because the mechanical properties of the capsules, permeability, interaction with the coating matrix and the transparency of the capsules can be tailored for the specific application.

2.2.2.2 Coacervation

Coacervation is a classical method for producing microcapsules and was used for instance for manufacturing the microcapsules for carbonless copy paper. The size of microcapsules produced by this method range from approximately 1 to several hundreds of microns. The principle for preparing microcapsules by coacervation is shown in Figure 2.20.

The process of coacervation follows basically three steps [13].

Step 1: Formation of three immiscible chemical phases

The immiscible phases are (i) a continuous phase, (ii) a core material phase, and (iii) a capsule material phase. To form these phases, a dispersion of the core material in a solution of the capsule material polymer in the continuous phase is formed (A in Figure 2.20). The capsule material phase is formed by precipitation of the capsule material (B in Figure 2.20). The precipitation can be started be utilising one of the following methods:

- Changing the temperature
- Adding a salt
- Adding a non-solvent
- Adding an incompatible polymer, inducing polymer-polymer interaction

The term **"complex coacervation"** is used for a coacervation process where the precipitation is induced by polymer-polymer interaction due to the usage of complementary polyelectrolyte materials. An example of complex coacervation is the classic microencapsulation process with the system gelatin-gum arabic (see below).

Step 2: Deposition of the capsule polymer on the core material

This step is carried out under controlled mixing. Deposition of the capsule polymer around the core material occurs if the polymer is absorbed at the interface between the core material and the continuous phase. The deposition (C in Figure 2.20) and subsequent coalescence of the capsule polymer (D in Figure 2.20) are promoted by a reduction of the total free interfacial energy of the system, brought about by decrease of the capsule material surface area.

Step 3: Rigidising the capsule

This is normally achieved by thermal, crosslinking or desolvation techniques, leading to microcapsules with suitable mechanical properties for the intended application.

One classical encapsulation procedure for dye precursors for the production of carbonless copy paper is carried out as follows:

"The solution of a gellable hydrophilic polymer gelatin is placed in a stirred vessel and the core material, e.g. the solution of dye precursor, is emulsified in it until the desired particle size of say 2 to 5 μm has been obtained. Gum arabic solution is then added. Coacervation occurs, the system being constantly stirred. Subsequently the composition is cooled from 50 °C to 5 to 10 °C and the gelatin-gum arabic complex coacervate which has been precipitated around the core phase is allowed to gel. It is hardened by addition of glutaraldehyde or formaldehyde and slow adjustment of the pH to 9 to 10 with caustic soda." (Cited from [1])

Other hydrophilic polymers that can be used as capsule materials are, for example, polyvinyl alcohol (PVA), methyl cellulose, and cellulose acetate. The coacervation technique is widely used for the microencapsulation of pharmaceutical agents.

2.2.2.3 *Mini-emulsion polymerisation*

The mini-emulsion technique can also be used to produce microcapsules. Oil in water emulsion is prepared in which the oil phase (containing the active agent) also contains a solved polymerisable monomer. The monomer and the oil have to be mixed before the polymerisation process starts. When the polymerisation starts, phase separation occurs and at the end of the polymerisation microcapsules are obtained which contain the active agent. By selecting an optimal dispersion technique and an appropriate surfactant, microcapsules with diameters down to 50 nm can be obtained [14].

2.2.2.4 *Spray drying*

Spray drying is a microencapsulation process that involves spraying an emulsion or dispersion into a stream of hot gas. The material to be sprayed contains a film-forming polymer that is dissolved in the continuous phase. The drying process leads to a firmly closed polymer capsule that encloses the core material. A very common application of this process is the production of dust-free pigment granules. The spray drying process can also be used to produce microcapsules that contain a liquid active agent [15]. Such microcapsules can have diameters down to 1 μm.

2.2.3 *Release of active agents*

In many cases the desired function given to the coating by the microcapsules is due to the release of the active agent from the microcapsules. There are different options here.

Controlled release through a semi-permeable capsule wall

In this case the capsule wall has to be permeable by the active agent. The porosity of a capsule wall can be tailored to the nature of the active agent and to the desired release rate. This is a very common approach in the pharmaceutical industry, where the adjustment of a defined release rate of drugs and pharmaceuticals is of decisive importance.

The morphology of the capsule wall can be influenced by the chemical composition and the morphology of the wall material. Methods have been described to equip the capsule walls with nano-sized holes or to even prepare cage-like microcapsules (see Figure 2.21) [16].

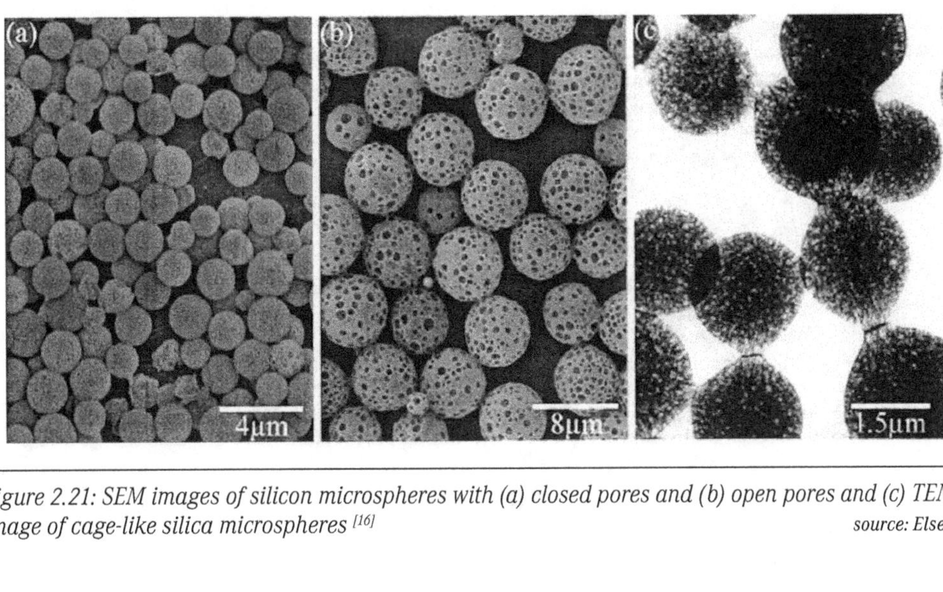

Figure 2.21: SEM images of silicon microspheres with (a) closed pores and (b) open pores and (c) TEM image of cage-like silica microspheres [16] source: Elsevier

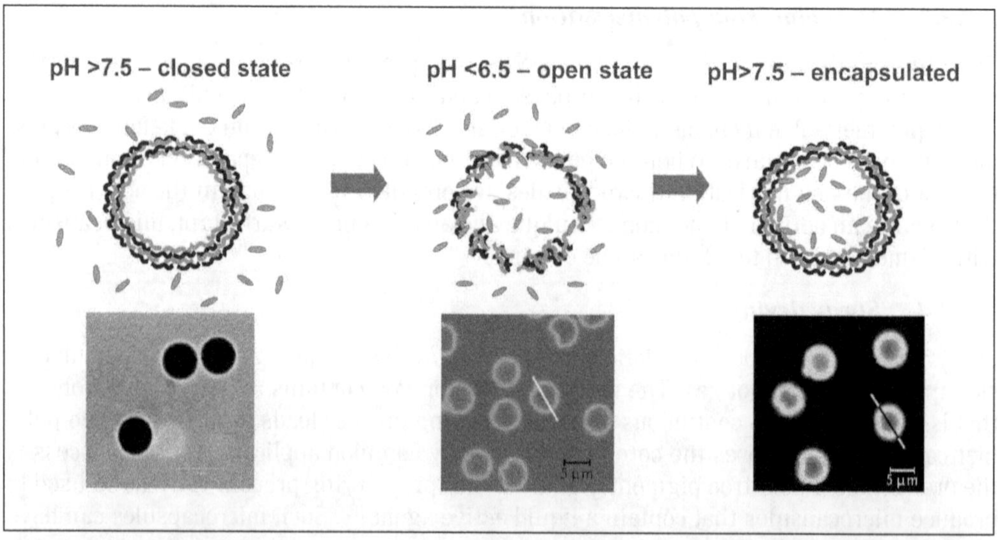

Figure 2.22: Microcapsules with pH-triggered opening and closing mechanism. The pictures below show the distribution of the fluorescent particles (rod-shaped particles in the scetch above) in a fluorescence-microscope image
source: H. Möhwald, MPI of Colloids and Interfaces

Triggered release via degradation of the capsule walls by environmental influences

A way to release the capsule content exactly when it is needed is via degradation of the capsule walls due to changes in environmental conditions. These changes include temperature, pH, UV light, and humidity. A variety of applications using this type of release concept are known from the pharmaceutical, agrochemical and food industries. Figure 2.22 shows an example of microcapsules that can be opened and even closed again by changing the pH-value of the environment.

Release due to mechanical damage to the microcapsules

Microcapsules that are used in coatings to achieve a self-repairing function are usually designed to be opened by rupture of the microcapsules. The same mechanism can be applied for the release of fragrances and lubricants from printing inks or coatings.

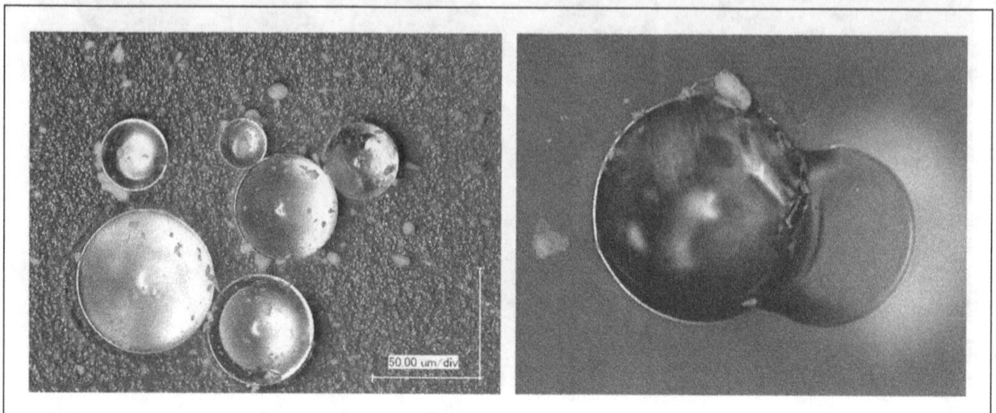

Figure 2.23: Left: Microcapsules filled with oxidative drying agent; Right: Punctured capsule with released content
source: Fraunhofer IFAM

For such applications the capsule walls have to be impermeable to the active agent. Usually the capsule walls have to protect the content and consequently have to be impermeable to moisture, oxygen, and other substances that could affect the encapsulated active agent. Figure 2.23 shows the release of a resin with low viscosity from a microcapsule that has been punctured.

Sudden release by nano-explosions

Recently a release mechanism has been reported that works via explosion of the microcapsules [17]. In this system the embedding of an azo-initiator that decomposes at a temperature below the T_g of the capsule polymer leads to sudden explosion of the capsule and immediate release of the active agent. The release temperature is defined by the decomposition temperature of the initiator. Other explosives can also be used for this purpose.

2.2.4 *Perspectives for using microcapsules in functional coatings*

The incorporation of microcapsules into polymeric films opens up interesting opportunities for the development of functional coatings. The coatings function is determined by the microencapsulated active agents. Several approaches for using encapsulated active agents have been reported:

- Lubricants for the preparation of anti-friction coatings [18]
  The encapsulated lubricant is released due to damage to the microcapsules caused by wear. The encapsulated lubricant can extend the lifetime of a heavy duty anti-friction coating significantly.
- Phase-change materials (PCMs) as heat-regulating agents [3]
  The microcapsules are filled with a material that melts in a given temperature range (e.g. around 20 °C). Typical materials are waxes or special salts. At temperatures above the melting point the PCM melts and absorbs heat (enthalpy of fusion). At temperatures below the melting point the PCM freezes again and gives the heat free. Such microcapsules can be incorporated into building paints or plaster or other construction materials as very energy-effective method of controlling the temperature.
- Fragrances for inks/lacquers [19]
  A major field for the application of microencapsulated fragrances is advertising in print media. For instance, advertisements for cosmetics companies in print media are sometimes equipped with microencapsulated fragrances that are released when the pictures are touched or rubbed.
- Polymerisable agents
  The usage of microencapsulated healing agents is a widely studied approach for self-repairing coatings. Such self-repairing systems often consist of microencapsulated, polymerisable agents, solvents or catalysts. Examples for polymerisable microencapsulated active agents are as follows:

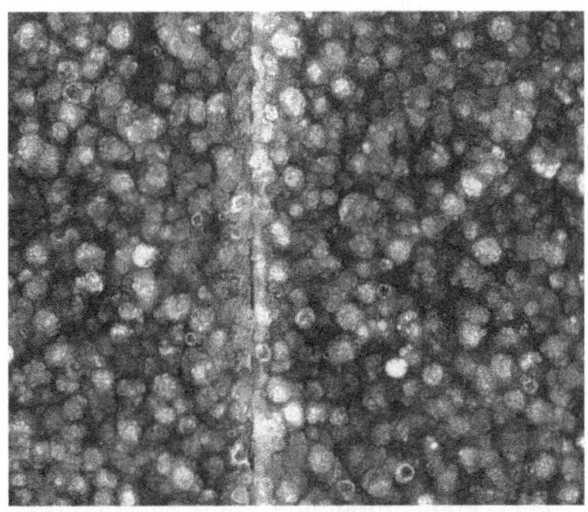

Figure 2.24: Microcapsules filled with fluorescent dye in a clear coat. The scratch leads to rupture of the microcapsules that are undamaged in the paint film. The clear coat was chosen only for investigation purposes. In a pigmented system only the scratch would be fluorescent source: IFAM-Fraunhofer

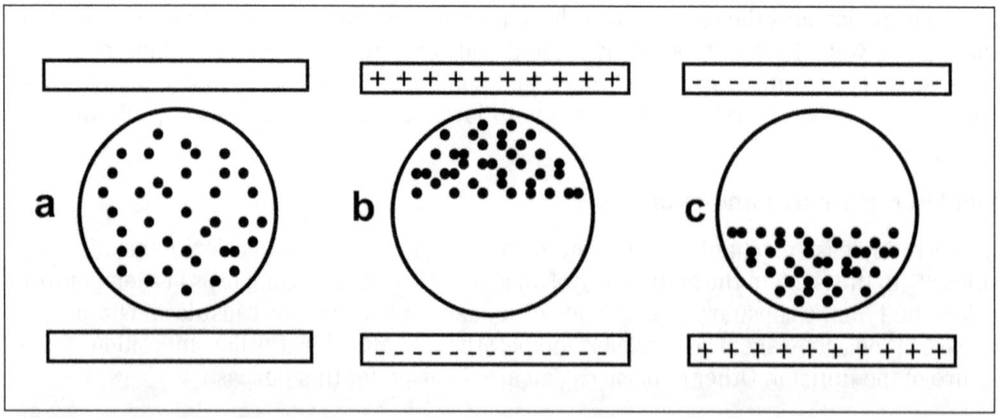

Figure 2.25: Electrophoretic colour-switching: a: no electric field, b and c: accumulation of nanoparticles due to the electric field

- Polythiols as crosslinkers for epoxy-based self-healing materials[20]
- Epoxy-resins as component for self-healing materials [21]
- Dicyclopentadiene (DCPD) for self-healing materials [22, 23]
- Isocyanates for self-healing polymers [24]

Information about self-repairing coatings is given in Chapter 3.3 "Self-healing coatings".

- Damage indicating coatings
  Microcapsules filled with dyes can very effectively indicate cracks in a coatings material. This can be an important function for coatings on basic materials where hits that lead to damages in the structure of the material are not obvious (e.g. fibre reinforced plastics). Figure 2.24 shows an example of a clear coat that is equipped with microcapsules containing a fluorescent dye. A clear coat was chosen for investigation purposes because it shows clearly what happens with the microcapsules and the dye. In the crack the capsules are ruptured and the dye is released, clearly indicating the damage. The force that is needed to rupture the capsules can be determined by the diameter and the wall thickness of the microcapsules.

- Electrophoretic or magnetic particle suspensions for colour-changing coatings
  Coatings that change colour on demand are of interest for the development of displays, electronic signs, and other products where switchable colour is useful. One approach is based on the microencapsulation of TiO_2 nanoparticles dispersed in a dielectric dye solution [25]. When such capsules are put in an electric field the nanoparticles accumulate on one side of the capsule due to their electrophoretic behaviour (zeta-potential) and dominate the colour of this side of the capsule. The colour of the other side is dominated by the dye solution (see Figure 2.25).

Such microcapsules can be switched between two colours depending on the dye and the colour of the nanoparticles. If colour-changing microcapsules are needed that can be switched between any colour, another highly sophisticated approach can be used based on photonic crystals embedded in microcapsules [26]. The content of the microcapsules (clusters of superparamagnetic Fe_3O_4 nanoparticles that are coated with polyacrylate) is solid at room temperature. When the capsules take up energy, the content becomes liquid and the colour can be tuned by an external magnetic field through the entire visible spectrum. When then the capsule content becomes solid again, the colour is frozen. This process is fully reversible and can be repeated (in theory) an unlimited number of times. Ongoing research projects are bringing this technology to industrial maturity [27].

2.2.5 Literature

[1] *Sliwka, W.*; Microencapsulation, Angew. Chem. Internat. Edit., 14, 1975, 539-550

[2] *Gouin, S.;* Microencapsulation: industrial appraisal of existing technologies and trends, Trends in Food Science & Technology 15, 2004, 330-347

[3] *Cho, J.-S.; Kwon, A.; Cho, C.-G.*, Microencapsulation of octadecane as a phase-change material by interfacial polymerization in an emulsion system, Colloid Polym. Sci. 280, 2002, 260-266

[4] *Gosh, S.K.* (edit.); Functional coatings by polymer microencapsulation, WILEY-VCH Verlag GmbH & Co. KGaA, Weinheim 2006

[5] *Blaiszil, B.J., Sottos, N.R., White, S.R.*, Nanocapsules for self-healing materials, Composites Science and Technology, 68, 2008, 978-986

[6] Ma, G.H.; Su, Z.G.; Omi, S., Sundberg, J.; Stubbs, J., Microencapsulation of oil with poly(styrene-N,N-dimethyaminoethylmethacrylate) by SPG emulsion technique: Effects of conversion and composition of oilm phase, J. Colloid Interface Sci., 266, 2, 2003, 282-294

[7] *Yuan, L.; Gu, A.; Liang, G.*, Preparation and properties of poly(urea-formaldehyde) microcapsules filled with epoxy resins, Mater. Chem. Phys. 110, 2008, 417-425

[8] *Brown, E.N.; Kessler, M.R., Sottos, N.R., White, S.R.*, In situ poly(urea-formaldehyde) microencapsulation of dicyclopentadiene, J. Microencapsulation, 20, No.6, 2003, 719-730

[9] *Hu, J.; Chen, H.Q.; Zhang, Z.*, Mechanical properties of melamine formaldehyde microcapsules for self-healing materials, Materials Chemistry and Physics, 118, 1, 2009, 63-70

[10] *Pascu, O.; Garcia-Valls, R. Giamberini, M.*, Interfacial polymerization of an epoxy resin and carbolxylic acids fort he synthesis of microcapsules, Polym. Int 57, 2008, 995-1006

[11] *Yadav, S.K.; Khilar, K.C.; Suresh, A.K.;* Release rates from semi-crystalline polymer microcapsules formed by interfacial polycondensation, Journal of Membrane Science 125, 1997, 213-218

[12] *Yeom, C.K.; Oh, S.B.; Rhim, J.W., Lee, M.J.*, Microencapsulation of water-soluble herbicide by interfacial reaction. I. Characterisation of microencapsulation, Journal of Applied Polymer Science, 78, 2000, 1645-1655

[13] Coacervate. (2009, July 29). In *Wikipedia, The Free Encyclopedia.* Retrieved 10:58, October 24, 2009 from http://en.wikipedia.org/w/index.php?title=Coacervate&oldid=304864420

[14] *Landfester, K.*, The generation of nanoparticles in miniemulsions, Advanced materials, 13,10, 2001, 765-798

[15] *Ninomya, Y.; Komamura, C.; Musa, Y.*, Verfahren zur Herstellung von Mikrokapseln, die ein flüssiges aktives Material enthalten, Patent DE3417200, 1985

[16] *He, X;, Shi*, X, Self-repairing coating for corrosion protection of aluminum alloys, Progress in Organic Coatings, 65, 2009, 37-43

[17] *Volz, M.; Walther, P.; Ziener, U.; Landfester, K.* Nano-explosions of nanoparticles for sudden release of substances by embedded azo-components as obtained via the miniemulsion process, Macromol. Mater. Eng., 292, 2007, 1237-1244

[18] *Capelli, A.*, US Patent 4056478, Bearing material employing frangible microcapsules containting lubricant, 1977

[19] Company brochure Follmann & Co. Gesellschaft für Chemie-Werkstoffe und -Verfahrenstechnik mbH & Co. KG, http://www.follmann.de

[20] Yuan, Y. C.; Rong, M. Z., Zhang, M. Q., Preparation and characterization of microencapsulated polythiol, Polymer 49, 2008, 2531-2541

[21] *Yuan, L.; Liang, G.; Xie, J.; Li, L.; Guo, J.*, Preparation and characterization of poly(urea-formaldehyde) microcapsules filled with epox resins, Polymer 47, 2006, 5338-5349

[22] *White, S.R.; Sottos, N.R.; Geubelle, P.H., Moore, J.S., Kessler, M.R., Sriram, S.R., Brown, E.N., Viswanathan, S.*, Autonomic healing of polymer composites, Nature, 409, 2001, 794-797

[23] *Stenzel, V.; Mock, U.; Tillner,* S., DIY repairs - Self-healing mechanisms are beeing developed for protective coatings, European Coatings Journal, 11, 2006, 32-36

[24] *Yang, M., Keller, A.W., Moore, J.S., White, S.R., Sottos, N.R.*, Microencapsulation of isocyanates for self-healing polymers, Macromolecules, 41(24) 2008, 9650-9655

[25] *Kim, K.S.; Lee, J.Y.; Park, B.J.; Sung, J.H.; Chin, I.; Choi, H.J.; Lee, J.H.*, Synthesis and characteristics of microcapsules containing electrophoretic particle suspensions, Colloid. Polym. Sci., 284, 2006, 813-816

[26] *Ge, J.; Hu, Y.; Yon, Y*, Highly tunable superparamagnetic colloidal photonic crystals, Angew. Chem. Int. Ed. 46, 2007, 7428-7431

[27] *Margutti, S.*, Chameleon surface, color change through microencapsulated photonic crystals, presentation of the Fraunhofer-Innovationszentrum, October 2009

2.3 Functional nanoparticles/nanofillers

Nanotechnology has been a buzzword over the last 15 years, and has become synonymous with high expectations. The impression was given that “nanomaterials” are something completely new and revolutionary from a technical standpoint. For paint chemists that has been difficult to understand, because nanoscale materials have been used in paints and coatings for a long time in order to give those materials specific properties. In the meantime the debate has moved on and has almost been turned around: In recent years the focus has increasingly been on potential risks. The danger now is that the discussion about risks shifts the emphasis away from the very promising developments in nanotechnology in recent years. The sections which follow focus on these technological developments.

The German government, just one of many funding bodies all over the world, funds research projects in the area of nanotechnology to the tune of ca. 330 million euros per year [1]. There are high expectations of these projects, some of which are specifically focused on coating materials. Figure 2.26 shows, for example, the expected applications of nanotechnology in the coming 10 to 15 years. The topics connected with the development of coating materials are highlighted in bold letters.

The field of “nanocoatings” or “nano-based coatings” has grown so much in the last 15 years that it is impossible to give a complete overview in a book like this. Even a summary of the nanoparticles or nanofillers used in coating systems would far exceed the confines of this book. For that reason, only nanomaterials of interest for the development of functional coatings are presented here, with example potential applications being given. A deeper insight in nanotechnology is given in another book in this series [2].

2.3.1 Nanotechnology – What does “nano” really mean?

A nanometer is a billionth of a meter (10^{-9} m). In common parlance the word nanotechnology refers to the manufacture, study, and understanding of materials or structures having a length of ca. 1 nm to 100 nm in at least one dimension. This order of magnitude is of particular interest because it determines the properties of a particle. At this order of magnitude, the physical, chemical, and biological properties of a material differ

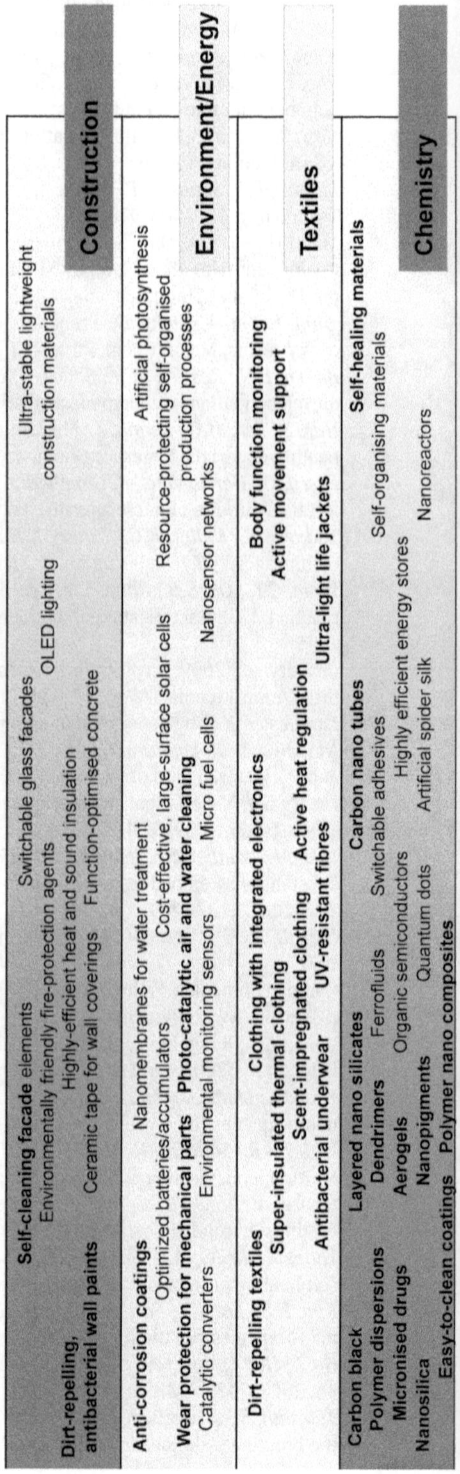

Figure 2.26: Examples applications and degree of maturity of nanotechnology in different sectors (reprinted

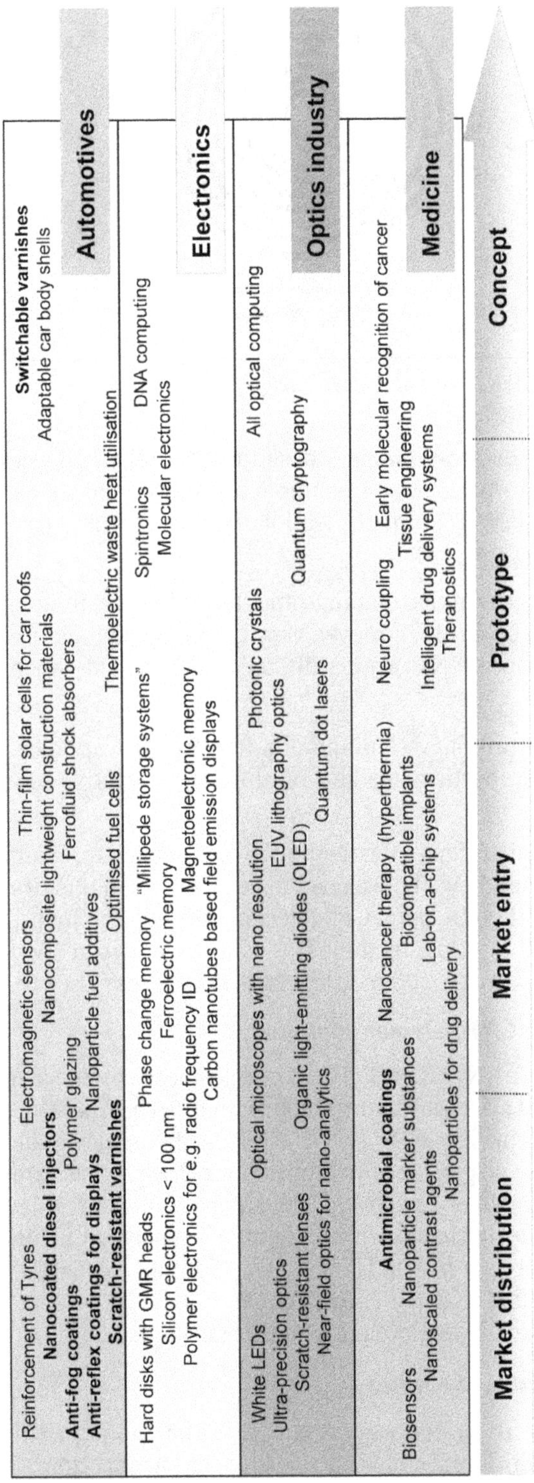

with permission of Dr. Baron, VDI TZ GmbH [1]. The items highlighted in bold letters by the authors are connected with coating technology.

fundamentally from the bulk properties of the same material. In the area of polymer nanocomposites, the extremely large area to mass ratio is of particular importance. Indeed, the properties of, for example, a nanoscale filler are largely determined by its surface properties. This fact allows coating materials to be developed which have new or much improved properties or functions.

2.3.2 Nanoparticles and nanocomposites – examples

2.3.2.1 Fibre-like nanoparticles (for example carbon nanotubes (CNTs))

Since their discovery in 1991, a wealth of research work has been carried out on carbon nanotubes (CNTs). Carbon nanotubes possess unique electrical and mechanical properties and are attractive for a host of applications:

- Young's modulus in the 100 to 1000 GPa range (for comparison, steel: 200 GPa, diamond: 1,200 GPa)
- Strength in the 2.5 to 63[3] GPa range (structural steel: 400 MPa, carbon fibre: 5.6 GPa, glass fibre: 4.7 GPa)
- Thermal conductivity up to 3000 W/m K (glass: 1.1 W/m K, copper: 401 W/m K)
- The electrical conductivity is highly dependent on the chemical structure. There are CNTs with the properties of semiconductors [4], some with the conductivity of metals (comparable with copper [5]) and even superconducting CNTs with a transition temperature of 5K [6].

CNTs consist of carbon atoms which are present in sp^2-hybridized form

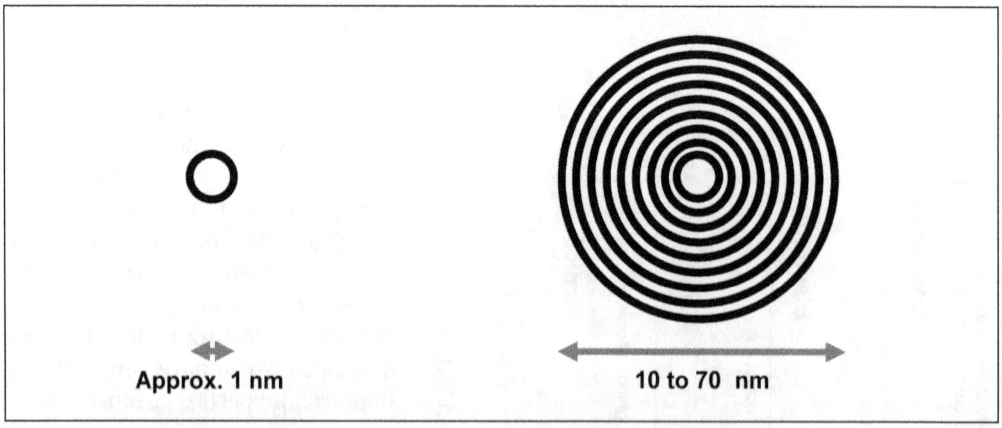

Figure 2.27: Schematic cross-section through a SWNT (left) and MWNT (right)

exactly as in graphite. In the simplest case one can imagine these tubes as rolled up graphite layers. A wide range of structures are possible, for example open or closed tubes, different diameters, single-wall nanotubes (SWNT) or multi-wall nanotubes (MWNT) (see Figure 2.27).

Single-wall nanotubes (SWNT) can have lengths in the μm range and these show the unique properties of CNTs in the purest way. These CNTs are, however, very tricky to manufacture and are extremely expensive. For this reason, SWNTs are hardly taken into consideration for industrial applications.

Multi-wall carbon nanotubes (MWNT) typically have lengths of a few microns up to ca. 50 μm. Due to their favourable cost and properties (see above), they are attractive raw materials for coatings.

Each of the three currently available methods for manufacturing CNTs (electric arc, laser, chemical vapour deposition (CVD)) is sufficiently well developed to be able to manufacture large quantities of CNTs of uniform quality (length, diameter, defects, walling). One industrial manufacturer is already producing 60 metric tons annually using a catalysed CVD process, and expansion of the capacity for a further 200 metric tons is in progress [7].

Modification of CNTs for manufacturing CNT-polymer composites

It can be expected from the properties of pure CNTs that CNT-polymer composites will have interesting technical properties (strength and electrical/thermal conductivity). The first CNT-polymer composite with an epoxy matrix was made back in 1994 [8]. CNTs tend to agglomerate because of their high surface energy. Effective dispersion and stabilisation of the particles are therefore vital when manufacturing a CNT-polymer composite. The surface of a CNT is relatively inert from a chemical viewpoint. For the mechanical strengthening of a polymer using CNTs, force transfer from the CNTs to the polymer matrix is imperative. For this reason, the surface of the CNTs must first of all be modified so that a force-fit connection to the polymer matrix can be made. A number of methods have been developed for this purpose:

Modifications of CNTs not involving covalent bonding

In this case the conjugated CNT – side wall system remains active and the physical properties are unaltered. The modification is possible via the physical absorption of polymers. The graphitic walls offer the opportunity for π-stacking interactions with conjugated polymers or for interactions with polymers which contain heteroatoms with free electron pairs [9].

Modifications of CNTs involving covalent bonding

This type of modification involves the formation of strong covalent bonds between the CNTs and the polymer matrix. These allow excellent force transfer. However, in these systems the conjugated CNT-wall system is disturbed, and this affects the physical properties of the particles. Here, the surface modification method must be carefully selected and/or adapted to give the desired properties of the composite.

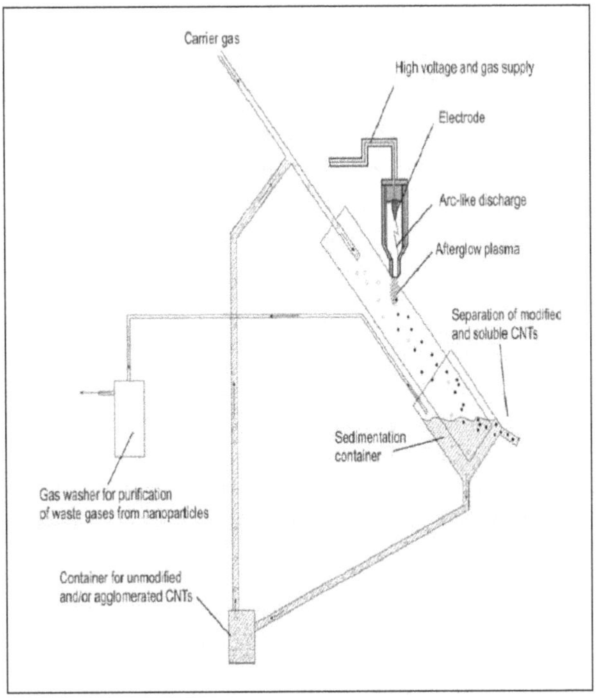

Figure 2.28: Device for oxidation of CNTs with atmospheric pressure plasma (patent pending) source: Fraunhofer IFAM

Oxidation of CNTs

Normally the first step of surface modification of CNTs is oxidation of the surface in order to generate oxygen-containing groups, and preferably carboxyl groups. On the one hand these improve the wetting properties of the surface and provide on the other hand points for chemical reactions with components of the polymer matrix. The oxidation can be carried out by wet-chemical means, for example using H_2O_2, HNO_3, or $KMnO_4$ [10]. Besides wet-chemical oxidation, a number of other methods are available for the oxidation via the gas phase. Possibilities here include oxidative powder treatment in a fluidised bed or oxidation of CNTs using an atmospheric pressure plasma method.

The latter can be directly coupled to the CNT manufacturing process or can be carried out in a separate processing step. Figure 2.28 schematically outlines such a process. Prior to the oxidation the CNTs are dispersed in water, or water is fed into the reaction chamber. The water allows a mild treatment with preservation of the structure of the CNTs (see Figure 2.29) and contributes to reducing the aggregate/agglomerate formation.

Chemical modification of oxidised CNTs

A fundamental distinction can be made here between the "grafting to" and the "grafting from" approach (see Chapter 2.5). The "grafting to" approach means that polymer chains

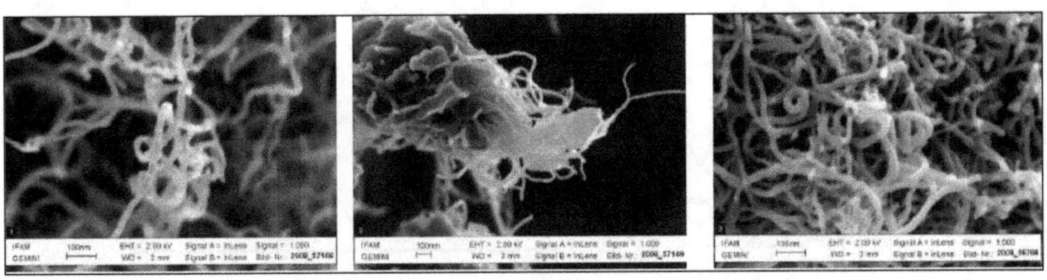

Figure 2.29: Unmodified CNTs (Baytubes C 150 P) (left); after dry atmospheric plasma treatment (center), and after atmospheric plasma treatment in the presence of water according to the process shown in Figure 2.28 (right) source: Fraunhofer IFAM

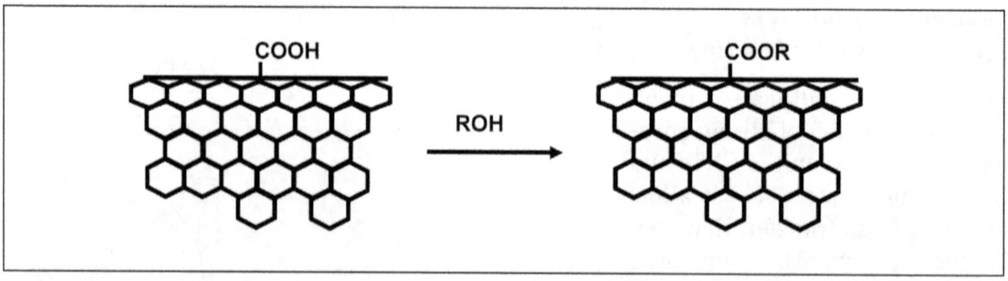

Figure 2.30: Ester bond between OH-groups on polymers and CNTs

diffuse to the functional groups and become attached there via chemical reactions. The advantage of this method is that commercially available polymers having defined properties can be used. The "grafting from" approach means that polymerisation occurs from initiator molecules attached to the functional groups of the CNTs. The advantage of this approach is that the realisable density of the modification on the surface of the CNTs is not limited, as it is with the "grafting-to" approach, by the steric hindrance of polymer chains which are already attached.

The carboxyl groups on oxidised CNTs allow a wealth of possible reactions with the functional groups of polymers. A few examples are shown below:

- Ester bond between oxidised CNTs and polymers [11, 12], see Figure 2.30

A similar reaction with amine groups on polymers leads to polymer modification via amide bonds.

- Nucleophilic substitution via organometallic bonds [13]

This mechanism requires prior reaction of the carboxyl groups with thionyl chloride ($SOCl_2$) in order to increase the reactivity via the formation of the acid chloride. Possible organometallic compounds include organic lithium compounds (Figure 2.31).

- Addition of isocyanate [14]

A variant of relevance for paints and coatings involves the reaction of the carboxyl groups with multifunctional isocyanates. The so modified CNTs can be reacted directly with polymers having OH groups and due to their covalent bonding to the polymer network they can be easily incorporated and have excellent force transfer properties.

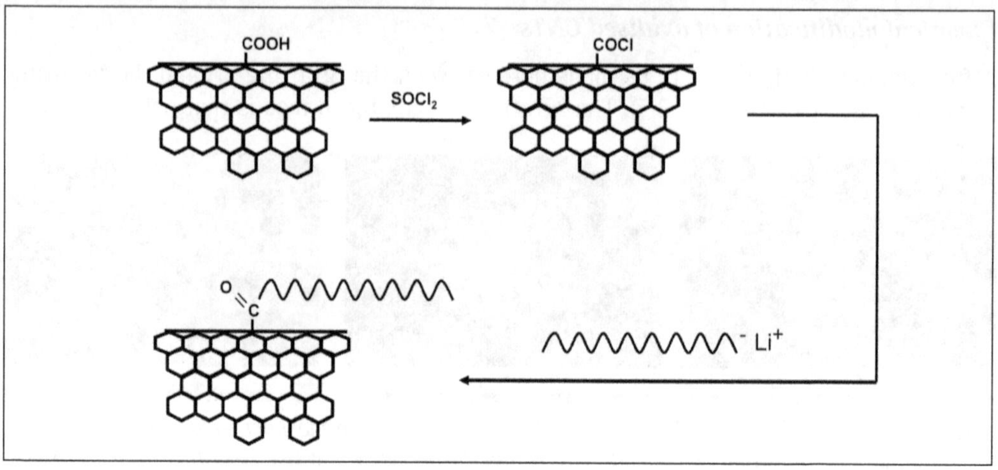

Figure 2.31: Nucleophilic substitution by organic lithium compounds

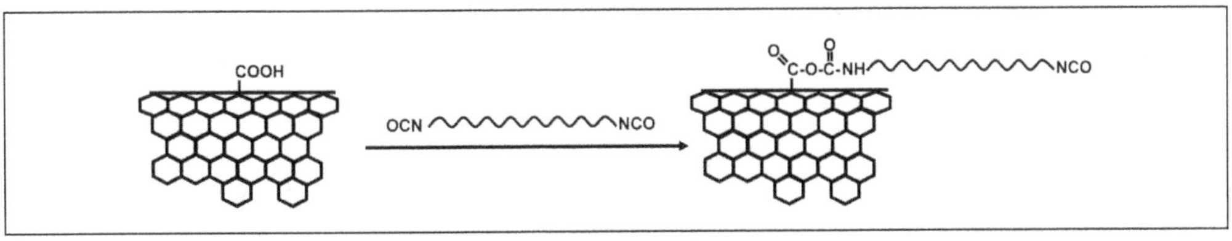

Figure 2.32: Reaction of oxidised CNTs with multifunctional isocyanates

In addition to the examples given here, there are scores of other reactions of oxidised CNTs described in the literature which can be used for the functionalisation of CNTs [9].

Besides the functionalisation of oxidized CNTs, modifications of non-pretreated CNTs with covalent bonds are also described. Compared to graphite, the stress caused by the "rolling up" of the graphite layers in CNTs leads to higher reactivity of the sp^2-hybridised carbon atoms. This state allows various reactions to be undertaken to modify CNTs for use in CNT-polymer composites. These reactions include, for example, the cycloaddition of azide-terminated polymer chains [15] and the reaction of hydroxy-functionalised primary amines with CNTs, which in turn offer a starting point for the subsequent polymerisation reactions, for example for the ring-opening polymerisation (ROP) with ε-caprolactone [16].

It can be stated in conclusion that a wide variety of chemical modifications of CNTs are possible. It is hence expected that suitable modifications of CNTs can be found for many applications of CNTs in coatings and paints.

Mechanical properties of CNT-polymer composites

The excellent strength of CNTs (see above) means that CNT-polymer composites should also have superior mechanical properties.

In order to achieve effective mechanical strengthening of polymers via the incorporation of CNTs, four system requirements must be met [17]:

- High aspect ratio of the CNTs (ratio of length to diameter)
- Good dispersion in the polymer matrix
- Alignment of the CNTs (less important, but random orientation is less effective than aligned CNTs)
- Force transfer between the CNT surface and the polymer matrix

For effective mechanical reinforcement of polymers, it is imperative to modify the CNT surface such

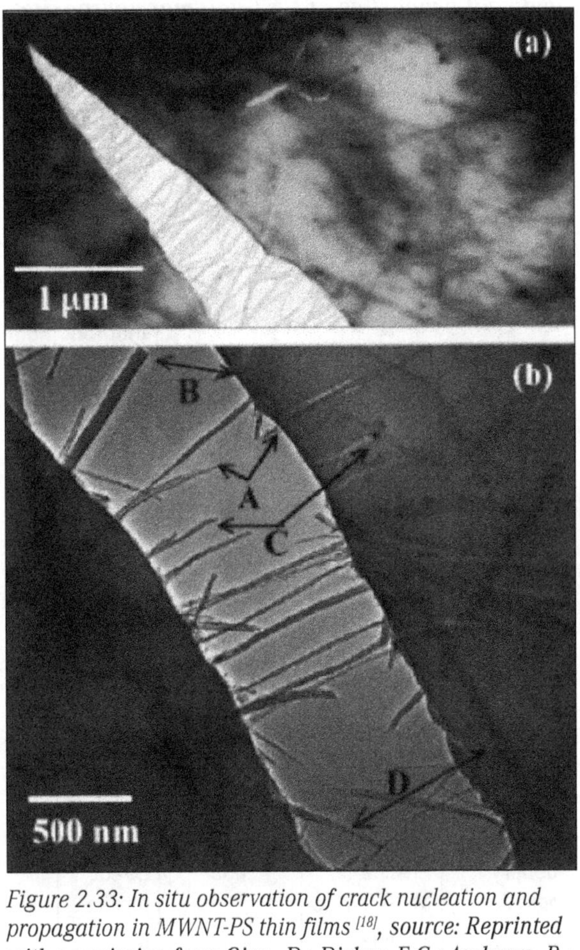

Figure 2.33: In situ observation of crack nucleation and propagation in MWNT-PS thin films [18], source: Reprinted with permission from Qian, D.; Dickey, E.C.; Andrews, R.; Rantell, T., Copyright 2010 American Institute of Physics

that there is effective attachment to the polymer matrix, so allowing good force transfer. In general, covalent bonds between CNTs and matrices are more effective for this than other types of interactions.

In recent years, literally hundreds of studies have been published about the reinforcement of polymers via the addition of CNTs. Some of the data, of special relevance for coatings, are presented here. In order to give an impression of the degree to which the mechanical properties of polymers can be improved by adding CNTs, some examples are given the following Table 2.2 [9].

Studies on the failure mechanisms of CNT-polymer composites have, for example, been conducted on polystyrene films reinforced with 1% CNTs [18]. These studies clearly showed that cracks are initially bridged by the CNTs (leading to mechanical reinforcement). However, beyond a certain crack width, or applied force, there is failure (see Figure 2.33). Either the CNTs themselves break or the CNTs are forced out of the polymer matrix. This illustrates that the attachment of the CNTs to the polymer matrix (which can be improved by pretreatment of the CNTs) is of key importance, but also that the strength of the CNTs themselves (related to structural defects) plays a role. The structural defects in turn are possibly caused by the oxidative treatment of the CNTs. The best possible compromise must hence be found here.

Electrical properties of CNT-polymer composites

CNTs have very high electrical and thermal conductivity (see above). Accordingly, it is to be expected that the use of CNTs in polymers results in composites which also have considerably higher electrical and thermal conductivities than the pure polymer materials.

Table 2.2: Mechanical properties of CNT-polymer composites

Matrix	CNT type	CNT weight fraction (%)	Processing method	Increase in tensile strength of the composite (%)	Increase in Young's modulus of the composite (%)	Ref.
PS	Purified MWNTs	5.00	Solution casting	50	120	[50]
PS	MWNTs, CPP-grafted	0.50	Solution mixing	100	100	[51]
Epoxy	Pristine MWNTs	3.00	Solution mixing	50	100	[52]
Epoxy	Copolymer modified MWNTs	0.25	Solvent mixing	50	50	[53]
Epoxy	MA-modified MWNTs	1.00	Solution mixing	45	100	[54]
Epoxy	Pristine MWTNs	0.50	Simple mixing	45	77	[55]
Phenolic resin	Dispersed MWNTs	3.00	Melt mixing	52	41	[56]
Rubbery epoxy	Pristine MWNTs	1.00	Direct mixing	100	100	[57]
Rubbery epoxy	Pristine MWNTs	4.00	Direct mixing	150	300	[57]

Table 2.3: Electrical properties of CNT-polymer composites

Matrix	CNT type	CNT weight fraction (%)	Processing method	Electrical conductivity of the composite (S/m)	Percolation threshold (weight %)	Ref.
PS	SWNTs	10	Solution mixing	6.89	0.045	[58]
PE	MWNTs,	10	Melt blending	10-3	7.5	[59]
PMMA	MWNTs	0.4	Solution mixing	$3x10^3$	0.003	[60]
PU	MWNTs	27	Solution casting	$2x10^3$	0.009	[61]
Epoxy	MWNTs	12	Solution mixing	$5x10^{-5}$	0.072	[62]
Epoxy	MWTNs	0.01	Shear mixing	2	0.0025	[19]
Epoxy	MWNTs	0.75	Roll milling	0.1	0.1	[63]
PVA	MWNTs	22.5	Solution mixing	0.27	0.72	[64]

The prerequisite for this is that the CNTs are effectively dispersed and stabilised in the polymer matrix, meaning that the individual nano-fibres form a conducting network within the polymer. The formation of a conducting network in turn requires that there is as close as possible contact between the particles. In this regard it is desirable for this application that the CNTs are not completely separated, but rather retain a certain degree of contact due to entanglements or aggregate formation. In general this is even guaranteed after being dispersed with high shear energy.

Studies on a 2-component epoxy-amine system have shown that with optimal incorporation of optimum multi-wall CNTs an extremely low percolation threshold (the percentage added at which a network is produced having significant conductivity - an interconnected network) can be achieved for about 0.0025 wt.% CNTs in the polymer [19]. By contrast, this threshold for spherical, conducting carbon black particles (tested in the same polymer matrix) is 10,000 times higher. The electrical conductivity in this system with 0.005 wt.% additive is already at an order of magnitude which permits the use of the polymer as an antistatic material (specific conductivity ca. 10^{-3} S/m) [19].

Many studies on modifying polymers with CNTs to increase the electrical conductivity have been carried out in recent years. The Table 2.3 gives a few examples to demonstrate the potential of CNTs for this purpose and shows the range of the results [9].

The data in the Table 2.3 clearly show that CNTs are very attractive for producing electrically conducting polymer composites, but that the specific properties depend on the choice of CNT type, the processing and incorporation method employed, and the choice of the polymer.

Besides giving excellent electrical conductivity, the addition of CNTs can also give polymers high thermal conductivity [20]. This is also an important aspect when using CNTs to manufacture functional coatings.

2.3.2.2 *Layered nanoparticles, for example nanoclays or layered double hydroxides (LDHs)*

Platelet-shaped fillers have long been used for controlling the rheology of paints and coatings (e.g. bentonite) and for improving specific properties such as the grindability or adhesion (e.g. talc). Platelet-shaped inorganic nanoparticles can in turn confer special properties on an organic paint film due to their shape and special chemistry. For example, the diffusion resistance can be considerably improved and fire-resistance properties can be conferred on the paint film.

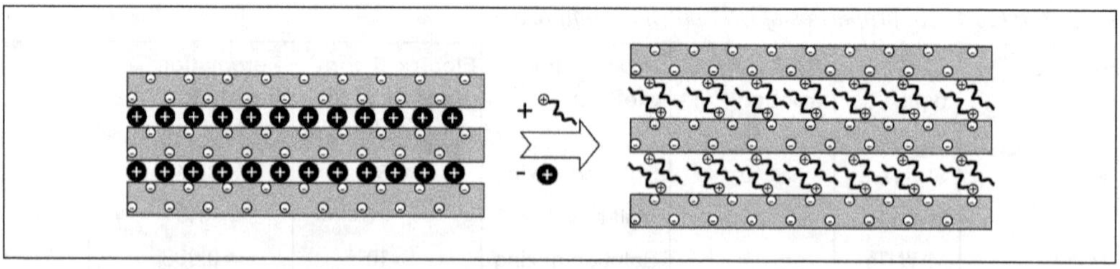

Figure 2.34: Organic modification of nanoclays

Clays are an important class of platelet-shaped, layered particles (e.g. bentonite, montmorillonite). They are crystalline and are composed of silicate layers and aluminium oxide layers. Clays, being inorganic particles, have poor intrinsic compatibility with organic binders. The individual layers of clay crystals are negatively charged. This negative charge is generally neutralised by the presence of sodium or calcium ions between the layers (Figure 2.34, left). Layered double hydroxides (LDHs) can to a certain extent be considered to be inverse clays. In these materials the layers are positively charged, whilst this charge is balanced by having inorganic anions between the layers.

A proven method for transforming the nanoclays into an organophilic form is to replace the inorganic cations by suitable organic cations (Figure 2.34, right). Quaternary ammonium ions are often used as organic cations. These often have a long organic chain which is compatible with standard binders.

Nanoclays are commercially available in modified forms like this. This exchange reaction naturally also offers the opportunity for further functionalisation of the layers with other agents.

One elegant method for modifying layer silicates with functional organic molecules is via a variant of so-called "click chemistry". This involves an exchange reaction with an organic quaternary ammonium salt (see before) followed by a further wet-chemical reaction to form azide groups, which can then be rapidly reacted with high conversion with various functional organic (and polymeric) molecules [21].

The next step in the manufacture of the nanocomposites is dispersion in the binder. In order to maximise the benefits of these fillers it is advantageous to separate the layers as much as possible in the paint film. If, for example, an 8 µm "nanoclay" particle is split into

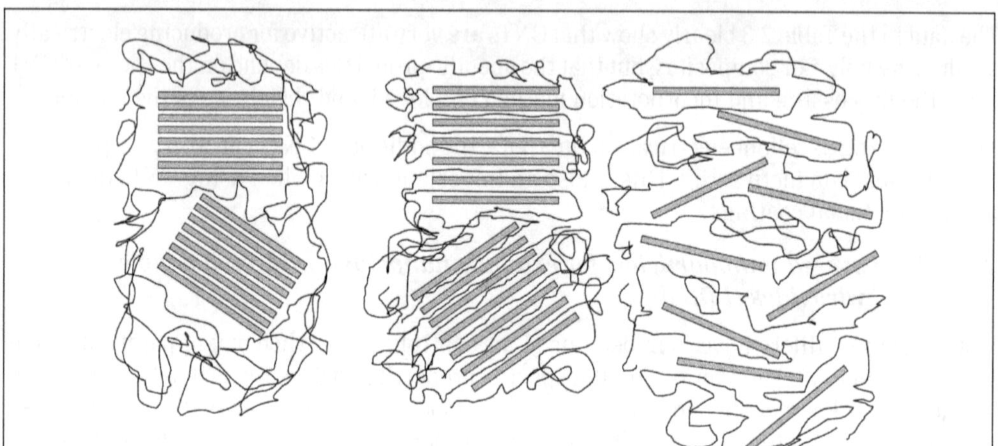

Figure 2.35: Different morphologies of layered nanoparticles in organic polymers

its layers (exfoliated), then the result is more than a million nanoplatelets [5]. Figure 2.35 shows the different possible morphologies of the particles in a binder.

The completely exfoliated state, in which the individual layers (having a thickness of ca. 1 nm) are completely separated and are surrounded by the binder, is the most advantageous and enables full utilisation of the benefits of these nanoparticles. In practice this is however only partially reached, as often there are still intercalated crystals, with binder between the layers but still in stacks. Figure 2.36 shows a typical micrograph.

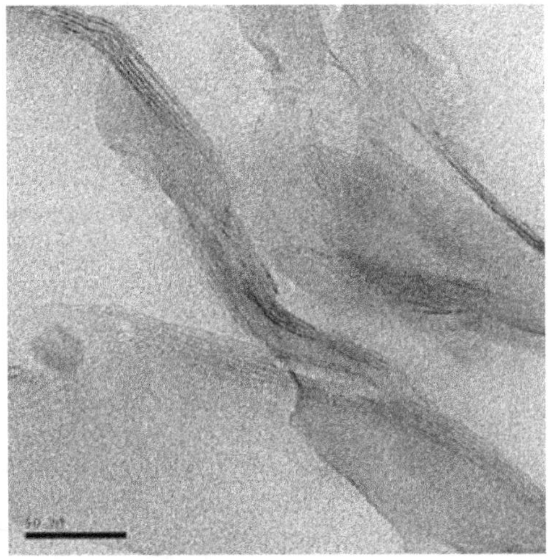

Figure 2.36: TEM image showing a typical mixture of intercalation and exfoliation (modified montmorillonite in an epoxy polymer) [22]
source: Schartel, B.; Knoll, U.; Hartwig, A.; Pütz, D, reprinted with permission of John Wiley & Sons Ltd.

Various methods can be used to manufacture such polymer nanocomposites:

- **Solution intercalation**
  Here the layer-structured nanoparticles are exfoliated in a solvent in which the polymers (or prepolymers for crosslinkable, thermoset systems) are dissolved. The solvent penetrates between the layers and leads to intercalation, and also to exfoliation if additional shear energy (e.g. using a bead mill, three roll mill, high speed mixer, ball mill) is introduced. This method can also be used for aqueous polymer solutions, and in this case the prior organophilic modification can be omitted. Examples of clay-polymer nanocomposites which can be manufactured using this method include: polyvinyl alcohol (PVA) [23], polyethylene oxide (PEO) [23], polyacrylic acid (PAA) [24], and polyimide (PI) [25].

- **Melt intercalation**
  Intercalation and exfoliation can also be carried out in a melt without any solvent. This method is naturally much used for the manufacture of thermoplastic polymer nanocomposites.

- **Emulsion polymerisation**
  Analogous to solution intercalation with water as a solvent, here the nanofiller is dispersed in water and subsequently subjected to an emulsions polymerisation step. This method can be employed, for example, for producing a PMMA-clay nanocomposite [26].

- **In-situ polymerisation**
  Liquid monomers which are intercalated between the layers can also polymerise there. This method has, for example, been studied in detail for the manufacture of epoxy-clay nanocomposites [27].

Effect of polymer nanocomposites based on layered particles on the properties of coating systems

The typical amounts of, for example, nanoclays which are used in coatings are ca. 2.5 to 5.0 wt.%. This amount, assuming there is optimum incorporation of the nanoparticles, allows certain properties of coatings to be significantly improved. A few examples are given below:

- In a typical amine-cured epoxy system the addition of 5 % modified montmorillonite (MMT) resulted in the following improvements [28]:

- Hardness: 8% (Buchholz method PN-EN ISO 2815; 2004)
- Scratch resistance: 55 % (PN-65/C-81527, Clemen scratch method)
- Abrasion resistance: 19 % (PN-76/C-81516, falling sand abrasion method)
- Water absorption: 46 % (PN-76/C-81521, gravimetric method)

• In extensive studies on the use of LDHs in a polyurethane-based automobile coating, the stone chipping resistance was considerably improved [29].
• Both clays [30] and also LDHs [31] have been successfully employed in UV curing coating systems, with property improvements comparable to those for the epoxy system above. The effect of the particles on the UV curing depends on the quality of the dispersion (the better the dispersion, the less the effect on the rate of curing) [32].
• In thermoplastic (thermally sprayable) coatings based on mixtures of ethylene vinyl acetate (EVA), ethylene vinyl alcohol (EVAI), and low density polyethylene (LDPE) the addition of 4 % nanoclay considerably improved both the mechanical properties (above all the abrasion resistance) and the corrosion protection properties. The latter is primarily due to the improved water vapour permeability due to the use of the nanoclays [33].
• Layer silicates significantly improve the fire protection properties of epoxy-based resins [34, 22]. If the ion exchange in the layer silicate (see above) involves phosphonium compounds instead of ammonium compounds, the fire protection properties of the layer silicate-polymer composites are improved even further.

2.3.2.3 *Spherical nanoparticles, for example nanosilica and other nanooxides*

The use of spherical nanoparticles in coating films has been the norm for many years. For the past 10 years or so nanoparticles have been used in automobile clearcoats to improve the scratch resistance. The most widely used particles here are pyrogenic silicas, and nanosilicas produced via the sol-gel method.

A good overview of the state-of-technology for the manufacture, surface modification, and use of nanoparticles for generating specific surface functions is given in another book in this series [2] and hence these aspects are only dealt with briefly here.

The following spherical nanoparticles are typically used for manufacturing functional coatings:

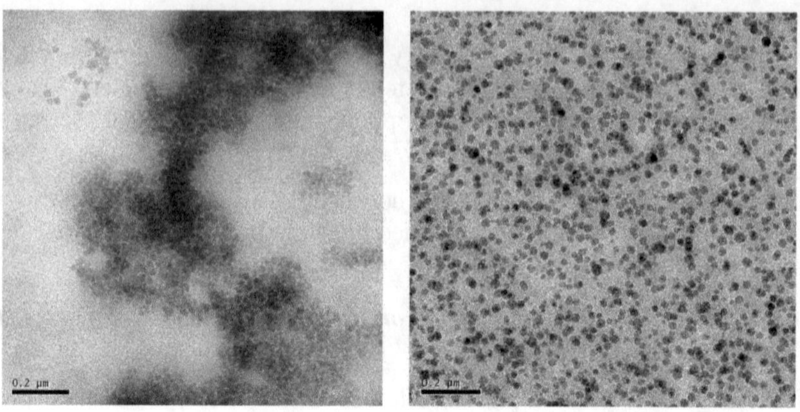

Figure 2.37: Left image: Flocculation of silica nanoparticles in a two-component polyurethane (PUR) coating due to the addition of hardener. Right image: Separated and stabilised nanoparticles in a cured film.
Source: Fraunhofer IFAM

- Pyrogenic silicas
  Improvement of the scratch resistance, corrosion protection
- Pyrogenic silicas with hydrophobic surface modification
  Hydrophobic coating surfaces, easy-to-clean coatings
- Silicas produced using the sol-gel method
  Improvement of the scratch resistance at high transparency
- Anatase (TiO_2)
  Photocatalytic activity for: Self-cleaning films, anti-fog layers, and anti-microbial layers
- Rutile (TiO_2)
  UV absorber, visual effects
- Zink oxide (ZnO)
  UV absorber, improvement of the scratch resistance, anti-bacterial properties [35]
- Corundum
  Improvement of the scratch resistance, abrasion resistance
- Indium-tin-oxide (ITO)
  Conducting layers, IR reflective coatings
- Magnetite (Fe_2O_3)
  Magnetic layers
- Ag "nanosilver"
  Anti-bacterial coatings

As is the case with other categories of nanoparticles, the biggest challenge for using these particles in organic coating systems is to guarantee good dispersion and stabilisation of the nanoparticles in the coating system. The difficulty here is due to the high surface area of the particles, which means there is a strong tendency towards flocculation. In order to manufacture suitable nanocomposites as starting materials for organic coatings, the same techniques are in principle used as for dispersing and stabilising pigments, albeit the requirements on the quality of composites manufactured with functional nanoparticles are often higher.

Flocculation must strictly be avoided when using nanoparticles in clear coats for applications where high requirements are put on the optical quality of the surface. This is because even a small fraction of agglomerates in the coating film leads to unacceptable opaqueness.

The tendency of nanoparticles to flocculate is often so great that the addition of standard surface-active additives (defoamers, wetting agents, slip-additives) or the addition of hardener in two-component systems can lead to the formation of aggregates.

An example of such behaviour is shown in Figure 2.37. The addition of an isocyanate hardener here to a polyol-based nanocomposite results in significant opaqueness in the clear coat system (left image). Only by employing an innovative stabilisation process was it possible to have the silica nanoparticles completely separated in the cured coating film and hence not the cause of light scattering. Such a coating film is completely transparent and is extremely scratch resistant and abrasion resistant due to the high content of hard nanoparticles (up to 25 wt.% in the solid coating).

2.3.2.4 Porous nanoparticles as hosts for active agents

A relatively new application of nanoparticles is their use as a reservoir for active agents which can confer special properties on a coating film. Such active agents include freezing

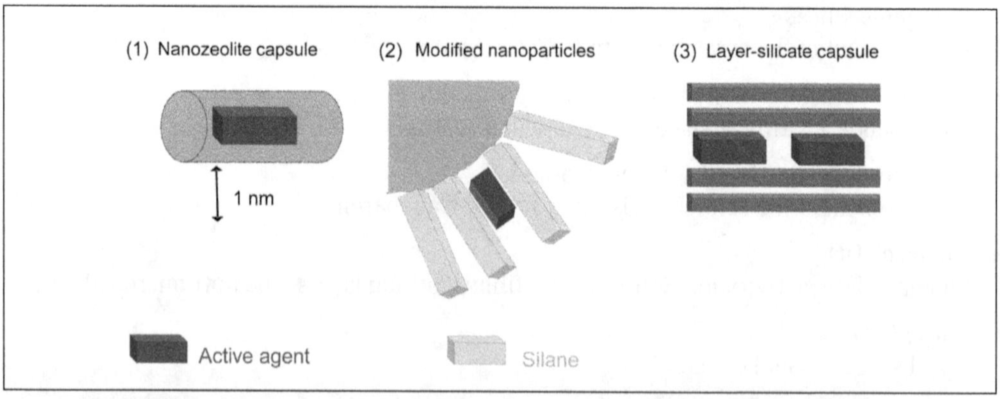

Figure 2.38: Different approaches for using nanoparticles as carriers of active agents in coating systems [36]

point lowering substances (see Chapter 3.2), corrosion inhibitors (see Chapter 3.3), and substances which significantly reduce the surface energy of the coating film and so give it "easy-to-clean" properties. Various types of nanoparticles can be used as carriers (see Figure 2.38):

- Nanozeolites (incorporation of active agents in the pores (see below))
- Layer silicates (intercalation of the active agents (see above))
- Spherical nanoparticles as carrier of active agents (adsorption on the surface, incorporation of the active agents into a layer of additives surrounding the nanoparticles, or grafting to suitable surrounding layers)

The use of such nanoparticles is a topic of ongoing research projects in the area of functional coatings. The use of nanoparticles will now be explained in more detail using the example of zeolites. Zeolites are crystalline aluminosilicate minerals whose framework structures are composed of AlO_4 and SiO_4 tetrahedra. The properties of zeolites are determined by their chemical composition, in particular the Si/Al ratio and other surface modifications. Typical features of zeolites are their uniformly shaped pores and/or channels, whose differing diameters (0.3 to 10 nm) make them suitable for a variety of applications. Zeolites, which are also known as molecular sieves, can therefore be used for separating chemical substances, for water purification, and also for heat storage and as drying agents. There is a huge range of naturally occurring and synthetic zeolites and reference should be made to, for example, McCusker et al. (2001) [37] or various Internet sites (www.iza-online.org or www.bza.org) for information about specific types of zeolites. Our main interest here concerns the benefits of using zeolites in functional coatings. So what are those benefits?

As already schematically illustrated in Figure 2.38, zeolites can adsorb molecules. These molecules can then be incorporated into coatings in a protected form and/or can provide additional surface functions. By specific control of the Si/Al ratio of the zeolite during the crystallisation process, optimum material properties can be created for the subsequent adsorption of guest species. A wide variety of examples of this can be found in the literature. Studies on the use of zeolites in conjunction with dye molecules as guest species have shown for example that the dye molecules are protected from chemical attack, photobleaching and thermal decomposition due to their incorporation in the zeolite. This allows new dyes to be developed and used. In one specific case zeolite L (see Figure 2.39) was loaded with dye molecules: cationic dyes were incorporated into the zeolite via ion exchange, and neutral dyes via gas phase adsorption. A feature of zeolite L is its large adsorption capacity; one cylindrically shaped crystal of 600 nm diameter has ca. 100,000 parallel channels

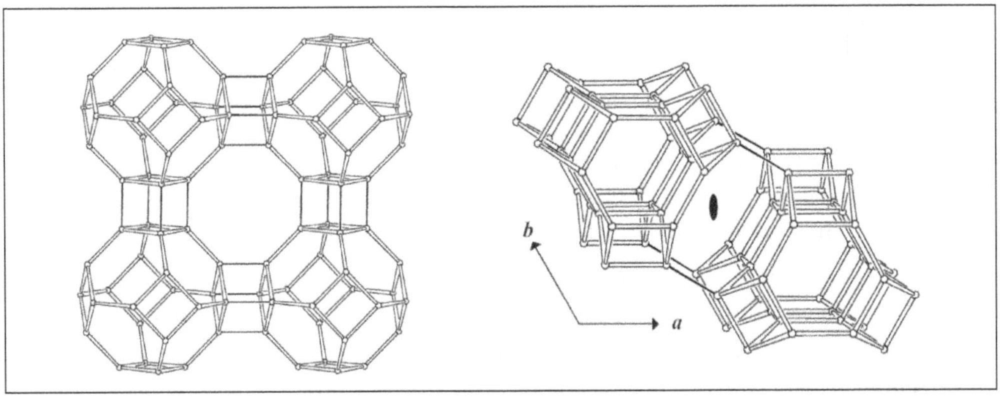

Figure 2.39: Structures of zeolite A (left) and zeolite L (right) respectively, source: H. van Koningsveld, Schemes for Building Zeolite Framework Models: http://www.iza-structure.org/databases/

which are able to adsorb many molecules. The loaded zeolites were then coated to ensure good dispersibility in the coating material. Applications for these materials were proposed for optical devices, for example lenses, special mirrors, and windows [38].

The use of zeolite-containing anti-fouling coatings has been investigated in a large number of studies. The focus up until now has been on zeolites filled with nano-silver. For example, condensing heat exchangers (condensers) were coated with zeolite A coatings for use on manned spacecraft. The deposition was carried out via in-situ crystallization and produced hydrophilic, polycrystalline coatings which were subsequently given anti-microbial functionality by ion exchange with silver salt solution [39]. A further example is the use of zeolite A loaded with 5 % silver (by weight) and 14 % zinc (by weight) which, when incorporated into an epoxy resin, provides protection for concrete sewer pipes against bio-corrosion [40].

Both naturally occurring and synthetic zeolites have been commercially available for many years. The question currently being addressed is how these highly porous inorganic materials can best be used in conjunction with organic coatings. As described above, proposals have already been put forward for anti-microbial coatings and for dye-loaded systems. Other potential uses of zeolites in coatings include for self-healing systems and as reservoirs for corrosion protection inhibitors. The challenges for future developments are:

- Selection of suitable types of zeolites
- Loading with suitable functional molecules
- Control of the release mechanisms (e.g. slow release, triggered release)
- Optimal dispersion in the organic coating

The matters to be addressed are very specific, meaning that tailor-made materials must be developed for specific applications. As in all development work, the relationship between the cost and benefit also plays a decisive role here [41].

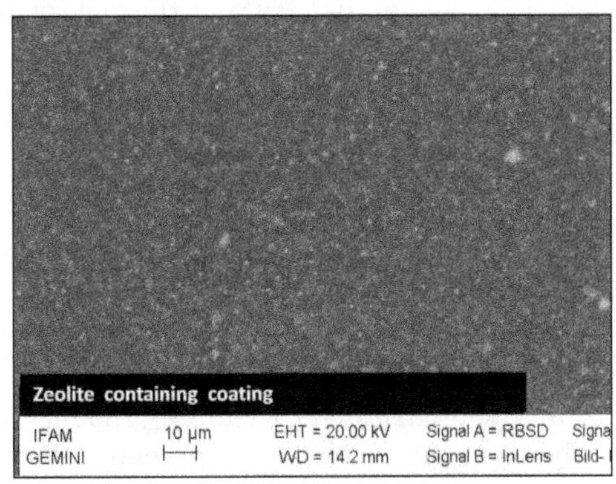

Figure 2.40: REM image of an epoxy-based coating containing nanozeolites (smaller than 1 µm)
source: Fraunhofer IFAM in cooperation with Nanoscape AG, Germany

2.3.3 *Safety, health, and environmental aspects*

Although this book focuses on new technologies, it is also appropriate here to mention any possible risks. The development of nanotechnology is way ahead of the understanding of the potential risk to people and the environment from the use of nanoparticles [42]. Given that, there is a certain risk that due to unexpected long-term effects certain nanoparticles will not be able to be used in the future. Besides any harmful effects on human health and the environment, this could also mean significant loss of invested capital.

In order to obviate these potential issues, the science of "nanotoxicology" has become established in recent years and is concerned with the potential risks of nanoparticles.

As the chemical and physical properties of nanomaterials can be completely different to bulk materials of the same composition due to the extremely high specific surface areas of the nanoparticles (gold, for example, is chemically very active in the form of nanoparticles), relatively little is known at present about the toxicology of many of these nanomaterials.

From recent studies it is known that certain nanoparticles can pass through membranes and penetrate into cells, tissue, and organs. This is not possible for larger particles [43]. Certain nanomaterials can get into the blood system by inhalation or oral uptake [44, 45], and there have even been reports about absorption through the skin [46].

Naturally, the fact that nanoparticles can get into the body and into organs does not necessarily mean that they cause harmful effects. This must be studied in detail for specific cases, and indeed there are numerous such studies currently underway [47].

Carbon nanotubes (CNTs) are at present very much the focus of toxicological studies because the shape (aspect ratio) of special sorts is comparable to the shape of asbestos fibres. As it is conceivable that special CNTs could be as harmful as asbestos, intensive work is underway. The harmful effects of asbestos, and in particular its carcinogenic effect, is primarily due to the size and shape of the asbestos particles (long, rigid fibres). It is suspected that long, rigid CNTs could have a similar effect. Animal studies in this area appear to confirm that. However, relatively short, less rigid CNTs show no tendency to cause cancer, and in particular not when they are present as entanglements [48]. Most of the currently commercially available CNTs belong to this group.

The estimation of the risk associated with using nanomaterials (often referred to as "engineered nanomaterials", ENM) is the topic of many ongoing research projects. For classifying the risks, a stepwise procedure was proposed which permits classification of the danger of nanomaterials with regard to hazard identification, hazard characterisation, exposure assessment, and risk characterisation [49].

The following can be concluded from the current status of the debate:

- Prior to significant investment in development projects which involve nanomaterials it is sensible to check the current status of the debate about the risks associated with nanoparticles.
- Nanoparticles which are incorporated into organic resins or polymers are in all probability non-hazardous, assuming the materials in bulk form are also non-toxic. As such, there appears to be no risk for end-users based on the current status of the debate.
- Concerning the handling of the relevant raw materials (the nanoparticles in dry form), the respective guidelines on work safety must be strictly observed.

2.3.4 Literature

[1] Federal Ministry of Education and Research (BMBF), Nano-Initiative - Action Plan 2010

[2] *Stefan Sepeur*, Nanotechnology - Technical Basics and Applications, Hanover: Vincentz Network 2008

[3] *Yu, M.; Lourie, O.; Dyer, M.J.; Kelly, T.F.; Ruoff, R.S.* Strength and breaking mechanism of multiwalled carbon nanotubes under tensile load, Science 287, 2000, 637-640

[4] *Kim, B.M; Fuhrer, A,M.S.*; Properties and applications og high-mobility semiconducting nanotubes, J. Phys: Condens Matter Vol. Issue 16, 18, 2004, 553-580

[5] *Koo, J.H.* Polymer nanocomposites: processing, characterization, and applications, McGraw Hill, 2006

[6] *Zang, Z.K.,; Zhang, L.Y.; Wang, N.; Zhang, X.X., Wen, G.H.; Li, G.D. et al.*, Superconductivity in 4 Angstrom single-walled carbon nanotubes, Science 292, no. 5526, 2001, 2462-2465

[7] Brochure Fa. Bayer Material Science, „The future is Nano""

[8] *Ayan, P.M.; Stephan, O., Colliex, C.; Trauth, D.*, Alligned carbon nanotube arrays formed by cutting a polymer resin-nanotube composite, Science 265, 1994, 1212-1214

[9] *Spitalky, Z. et al.*, Carbon nanotube-polymer composites: Chemistry, processing, mechanical and electrical properties, Prog. Polym. Sci., 2009, doi: 10.1016/j.progpolymsci.2009.09.003

[10] *Li, Y.-H.; Wang, S.; Luan, Z.; Ding, J.; Xu, C.; W, D.*, Adsorption of cadmium(II) from aqueous solution by surface oxidized carbon nanotubes, Carbon, Vol. 41, Issue 5, 2003, 1057-1062

[11] *Hill, D.E.; Lin, Y.; Rao, A.M.; Allard, L.F.; Sun, Y.P.*, Functionalization of carbon nanotubes with polystyrene, Macromolecules, 35, 2002, 9466-9471

[12] *Baskaran, D.; Mays, J.W.; Bratcher, M.S.*; Polymer adsorption in the grafting reactions of hydroxyl terminal polymers with multi-walled carbon nanotubes, Polymer, 46, 2005, 5050-5057

[13] *Huang, H.N.; Liu, I.C.; Chang, C.Y.; Tsai, H.C.; Hsu, C.H.; Tsiang, R.C.C.* , Preparing a polystyrene-functionalized multiple-walled carbon nanotubes via covalently linking acyl chloride funcionalities with living polystyryllithium, J. Polym. Sci. A Polym. Chem. 42, 2004, 5802-5810

[14] *Kuan, H.C.; Ma, C.C.M; Chang, W.P.; Yuen, S.M.; Wu, H.H.; Lee, T.M.*; Synthesis, thermal, mechanical and rheological properties of multiwall carbon nanotube/waterborne polyurethane nanocomposite, Sci. Technol., 65, 2005, 1703-10

[15] *Qin, S.; Qin, D.; Ford, W.T., Resasco, D.E.; Herrera, J.E.*, Functionalization of single-walled carbon nanotubes with polystyrene via grafting to and grafting from methods, Macromolecules, 37, 2004, 752-757

[16] *Buffa, F.; Hu, H.; Resasco, D.E.*, Side-wall functionalization of single-walled carbon nanotubes with 4-hydroxymethylaniline followed by polymerization of ε-caprolactone, Macromolecules, 38, 2005, 8258-8263

[17] *Coleman, J.N.; Khan, U.; Blau, W.J.; Gun'ko, Y.K.*, Small but strong: A review of the mechanical properties of carbon nanotube-polymer compostes, Carbon 44, 2006, 1624-1652

[18] *Qian, D.; Dickey, E.C.; Andrews, R.; Rantell, T.*, Load transfer and deformation mechanisms in carbon nanotube-polystyrene composites, Appl. Phys. Lett., 76, (20), 2000, 2868-2870

[19] *Sandler, J.K.W.; Kirk, J.E.; Kinloch, I.A.; Shaffer, M.S.P.; Windle, A.H.*, Ultro-low electrical percolation threshold in carbon-nanotube-epoxy composites, Polymer, 44, 2003, 5893-5899

[20] *Nan, C.-W.; Shi, Z.; Lin, Y.*, A simple model for thermal conductivity of carbon nanotube-based composites, Chemical Physics Letters, 375, 2003, 666-669

[21] *Oral, A.; Tasdelen, M.A.; Demirel, A.L., Yagci, Y.*, Poly(methyl methacrylate)/clay nanocomposites by photoinitiated free radical polymerization using intercalated monomer, Polymer 50, 2009, 3905-3910

[22] *Schartel, B.; Knoll, U.; Hartwig, A.; Pütz, D.*, Phosphonium-modified layered silicate epoxy resins nanocomposites and their combinations with ATH and organo-phosphorous fire redardants, Polym. Adv. Technol, 17, 2006, 281-293

[23] *Ogata, N.; Kawakage, S; Orgihara, T.*, Poly (vinyl-alcohol)-clay and poly (ethylene oxide)-clay blend prepared using water as solvent, J. Appl. Poly. Sci., 66, 1997, 573-581

[24] *Billingham, J.; Breen, C.; Yarwood, J.*, Adorption of polyamide, polyacrylic acid and polyethylene glycol on montmorillonite: An in-situ study using ATR-FTIR, Vibr. Spectrosc. 14, 1997, 19-34

[25] *Yano, K.; Usuki, A.; Okada, A.; Kurauchi, T., Kamigaito, O.*, Synthesis and properties of polyimide-clay hybrid, J. Polym. Sci. Part A: Polym. Chem., 31, 1993, 2493-2498

[26] *Lee, D.C.; Jang, L.W.*, Preparation and characterization of PMMA-clay hybrid composite by emulsion polymerization, J. Appl. Polym. Sci., 62, 1996, 1117-1122

[27] *Messersmith, P.B.; Giannelis, E.P.*, Synthesis and characterization of layered silicate-epoxy nanocomposites, Chem. Mater., 6, 1994, 1719-1725

[28] *Kowalczyk, K. Spychaj, T.*, Epoxy coatings with modified montmorillonites, Progress in Irganic Coating, 62, 2008, 425-429

[29] *Hintze-Bruening, H.; Troutier-Thuilliez, A.-L., Leroux, F.*, Layered particle based polymer composites for coatings: Part II-Stone chip resistant automotive coatings, Progress in Organic Coatings, 64, 2009, 193-204

[30] *Ceccia, S.; Turcato, E.A.; Maffetone, P.L.; Bongiovanni, R.*, Nanocomposite UV-cured coatings: Organoclay intercalation by an epoxy resin, Progress in Organic Coatings, 63, 2008, 110-115

[31] *Shichang, L, Zhou, W.; Miao, H.; Shi, W.*, Preparation and properties of poymer/LDH nanocomposite used for UV curing coatings, Progress in Organic Coatings, 65, 2009, 450-456

[32] *Landry, V.; Riedl, B.; Blanchet, P.*, Nanoclay dispersion effects on UV coating curing, Progress in Organic Coatings 62, 2008, 400-408

[33] *Tambe, S.P.; Naik, R.S.; Singh, S.K.; Patri, M.; Kumar, D.*, Studies on effect of nanoclay on the properties of thermally sprayable EVA and EVAl coatings, Progress in Organic Coatings, 65, 2009, 484-489

[34] *Hartwig, A.; Pütz, D.; Schartel, B.; Bartholomai, M.; Wendschuh-Josties, M*, Combustion behaviour of epoxide based nanocomposites with ammonium and phosphonium bentonites, Macromol. Chem. Phys, 204, 2003, 2247-2257

[35] *Li, J.H.; Hong, R.Y.; Li, M.Y.; Li, H.Z.; Zheng, Y.; Ding, J.*, Effects of ZnO nanoparticles on the mechanical and antibacterial properties of polyurethane coatings, Progress in Organic Coatings, 64, 2009, 504-509

[36] www.nanomodule.de

[37] *McCusker, L.B., Liebau, F., Engelhardt, G.*, Nomenclatur of structural and compositional characteristics of ordered microporous and mesoporous materials with inorganic hosts, Pure Appl. Chem. 73 (2), 2001, 381-394

[38] *Suárez, S., Devaux, A., Bañuelos, J., Bossart, O., Kunzmann, A., Calzaferri, G.*, Transparent Zeolite-Polymer Hybrid Materials with Adaptable Properties, Adv. Funct. Mater., 17, 2007, 2298-2306

[39] *McDonnell, A.M.P., Beving, D., Wang, A., Chen, W., Yan, Y.*, Hydrophilic and Antimicrobial Zeolite Coatings for Gravity-Independent Water Separation, Adv. Funct. Mater., 15 (2), 2005, 336-340

[40] *Haile, T., Nakhla, G., Allouche, E., Vaidya, S.*, Evaluation of the bactericidal characteristics of nano-copper oxide or functionalized zeolite coating for bio-corrosion control in concrete sewer pipes, Corrosion Science, 52, 2010, 45-53

[41] Dr. Wayne Daniell, NanoScape AG, Germany; personal communication

[42] *Durnev, A.D.*; Toxicology of nanoparticles, Bulletin of Experimental Biology and Medicine, 145, 2008, 72-74

[43] *Holsapple, M.P.; Farland, W.H.; Landry, T.D.; Monteiro-Riviere, N.A.; Carter, J.M.; Walker, N.J.; Thomas, K.V.*, Research strategies for safety evaluation of nanomaterials, part II: Toxicology and safety evaluation of nanomaterials, current challenges and data needs, Toxicological Sciences 88, 2005, 12-17

[44] *Oberdörster, G. et al.*, Principles for characterizing the potential human health effects from exposure to nanomaterials: Elements of a screening strategy, Particle and Fibre Toxicology 2, 2005

[45] *Hoet, P.H.M.; Brüske-Hohlfeld, I.; Salata, O.V., Nanoparticles* - known and unknown healt risks, Journal of Nanobiotechnology 2, 2004

[46] *Ryman-Rasmussen, J.P., Riviere, J.E.; Monteiro-Riviere*, Penetration of intact skin by quantum dots with diverse physicochemical properties, Toxicological Sciences 91, 2004, 159-165

[47] *Searon, A.; Tran, L.; Aitken, R.; Donaldson, K.*, Nanoparticles, human health hazard and regulation, J. R. Soc. Interface, 2009, doi: 10.1098/rsif.2009.0252.focus

[48] *Poland, C.A.; Duffin, R.; Kinloch, I.; Maynard, A.; Wallace, W.A.H.; Seaton, A.; Stone, V.; Brown, S.; MacNee, W.; Donaldson, K.*, Carbon nanotubes introduced into the abdominal cavity of mice show asbestos-like pathogenicity in a pilot study, Nature Nanotechnology, 3, 2008, 423-428

[49] *Savolainen, K.; et.al.*, Risk assessment of engineered nanomaterials and nanotechnologies – A review, Toxicology 2010, doi:10.1016/j.tox2010.01.013

[50] *Safadi, B.; Andrews, R.; Grulke, E.A.*, Multiwalled carbon nanotube polymer composites: Synthesis and characterization of thin films, J. Appl. Polym. Sci, 84, 2002, 2660-2669

[51] *Blake, R.; Coleman, J.N.; Byrne, M.T.; McCarthy, J.E.; Perova, T.S.; Blau, W.J., et.al*, Reinforcement of poly(vinyl chloride) and polystyrene using chlorinated polypropylene grafted carbon nanotubes, J. Mater. Chem. 16, 2006, 4206-4213

[52] *Park, J.M.; Kim, D.S.; Lee, J.R.; Kim, T.W.*, Nondestruktive damage sensitivity and reinforcing effect of CNT/epoxy composites using electro-micromechanical technique, Mater. Sci. Eng. C, 23, 2003, 971-975

[53] *Li, Q.; Zaiser, M.; Koutsos, V.*, CNT/epoxy resin composites using a block copolymer as dispersing agent, Phys. Stat. Sol. A 201, 2004, 89.91

[54] *Tseng, C.H.; Wang, C.C.; Chen, .Y.*, Functionalizing carbon nanotubes by plasma modification fort he preparation of covalent-integrated epoxy composites, Chem. Mater. 19, 2007, 308-315

[55] *Baj, J.B.; Allaoui, A.*; Effect of the length and the aggregate size of MWCNTs on the improvement efficiency of the mechanical and electrical properties of nanocomposites-experimental investigation, Compos. A 34, 2003, 689-94

[56] *Tai, N.H.; Yeh, M.K.; Liu, J.H.*, Enhancement of the mechanical properties of CNT/phenolic composites using a carbon nanotube network as the reinforcement, Carbon 42, 2004, 2774-2777

[57] *Allaoui, A.; Bai, S.; Cheng, H.M.; Bai, J.*, Mechanical and electrical properties of a MWNT/epoxy composite, Compos. Sci. Technol. 62, 2002, 1993-1998

[58] *Ramasubramaniam, R.; Chen, J.; Liu, H.*, Homogeneous carbon nanotube/polymer composites for electrical applications, Appl. Phys. Lett., 83, 2003, 2928-2930

[59] *McNally, T.; Pötschke, P.; Halley, P.; Murphy, M.; Martin, D.; Bell, S. et al*, Polyethylene multiwalled carbon nanotube composites, Polymer, 46, 2005, 8222-8232

[60] *Kim, H.M.; Kim, K.; Lee, S.J.; Joo, J.; Yoon, H.S.; Cho, S.J., et al*, Charge transport properties of composites of multiwalled carbon nanotube with metal catalyst and polymer: application to electromagnetic interference shielding, Curr. Appl. Phys., 4, 2004, 577-580

[61] *Koerner, H., Liu, W. Alexander, M.; Mirau, P.; Dowty, H.; Vaia, R.H.*, Deformation-morphology correlations in electrically conductive CNT-thermoplastic polyurethane nanocomposites, Polymer, 46, 2005, 4405-4420

[62] *Cui, S.; Canet, R.; Derre, A.; Couzi, M.; Delhaes, P.*, Characterization of multiwall carbon nanotubes and influence of surfactant in the nanocomposite processing, Carbon, 41, 2003, 797-809

[63] *Thostenson, E.T.; Ziaee, S.; Chou, T.W.*, Processing and electrical properties of caron nanotube/vinyl ester nanocomposites, Compos. Sci. Technol., 69, 2009, 801-805

[64] *Hernandez, Y.R.; Gryson, A.; Blighe, F.M.; Cadek, M.; Nicolosi, V.; Blau, W.J.*, et al, Comparison of carbon nanotubes and nanodiscs as percolative fillers in elecrically conductive composites, Scripta Mater. 58, 2008, 69-72

2.4 Sol-gel process

The sol-gel process is nowadays an established technology in the coating industry. Whereas classical paint chemistry uses organic components, the sol-gel process involves inorganic materials (typically silicon compounds). Coatings manufactured using the sol-gel process have different properties to those of classical coatings. For example, they have high hardness and excellent resistance to chemicals. Hybrid layers can also be generated by incorporating organic components into these systems, thereby uniting the properties of organic and inorganic coatings. Sol-gel coatings hence offer new opportunities for developing functional surfaces and for this reason we will introduce these systems here. For more detailed information please refer to the book by S. Sepeur (2008) [1] in this series.

2.4.1 Principle of the sol-gel process

The sol-gel process generally involves two reaction steps. Alkoxides of metals or metalloids are the starting materials (also termed precursors). There is first of all a hydrolysis reaction, as schematically shown in Figure 2.41 for the example of tetraethoxysilane (TEOS). This generates reactive hydroxyl groups. These then react with each other in a condensation reaction, liberating water, to form a network.

Besides the use of TEOS (as shown in Figure 2.41), other metal alkoxides such as titanium or aluminum alkoxide can also be used in this process. In this case pure inorganic network structures form. Initially there is formation of colloidal oligomers (so-called sols) and as the reaction proceeds these in turn react to form 3-dimensional networks having gel properties. All these reactions can be controlled by altering the medium, the concentration, the pH, and the temperature [1].

There are two reasons why the sol-gel process is important for developing functional surfaces: Firstly, it can be used to manufacture nanoparticles and secondly it can be used to manufacture coating films. It is clear that in order to produce nanoparticles the condensa-

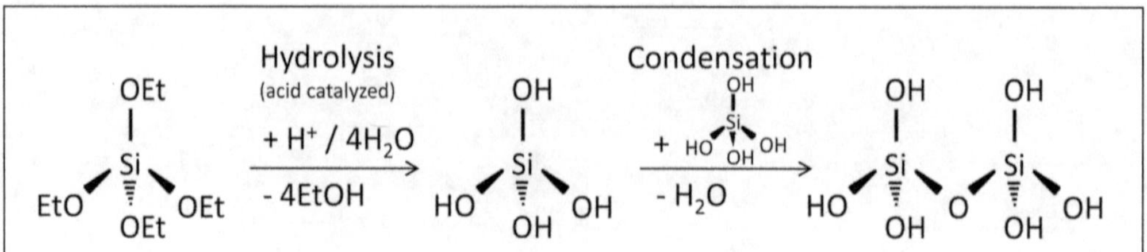

Figure 2.41: Schematic representation of the course of the classical sol-gel process with tetraethoxysilane (TEOS) as the example precursor source: [1]

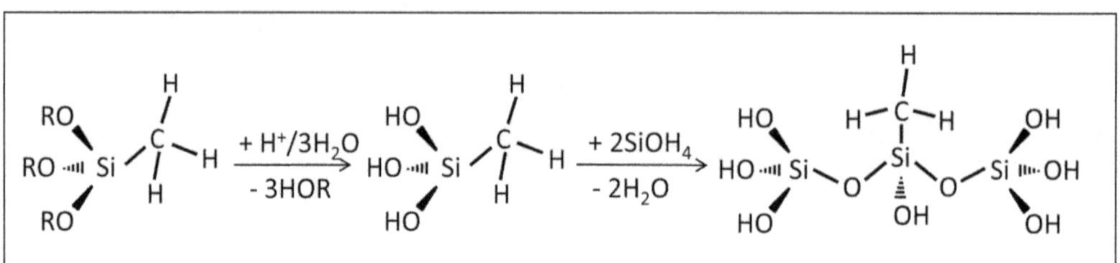

Figure 2.42: Schematic representation of network formation using methyltrialkoxysilane and production of a hybrid sol-gel with inorganic and organic components source: [1]

tion reactions must be stopped at the desired particle size. For example, nanoparticles can be produced from TEOS under suitable reaction conditions using aqueous ammonia in an alcoholic solution (bottom-up method) [1]. The applications of nanoparticles have been described in Chapter 2.3 and elsewhere. For example, nanosilicas produced via sol-gel synthesis are used to produce anti-scratch clear coats for the car industry.

In contrast, the formation of stable 3-dimensional networks is necessary in order to produce sol-gel layers and coatings. In the case of layers based on TEOS, the gels that are formed are sintered at high temperature to form high strength ceramic layers. Examples of sol-gel systems as functional coatings are described in the next section.

2.4.2 *Manufacture of functional coatings*

In addition to the pure inorganic coatings that have already been described, the sol-gel process can also be used to produce hybrid polymers containing both inorganic and organic components. This is due to the fact that silicon atoms can form covalent bonds with organic side chains. Figure 2.42 shows this for the example of methyltrialkoxysilane. Different organic groups can be used, so allowing a wide range of coating properties to be produced.

The combination of inorganic networks and organic functional groups offers great scope for producing functional coatings. For example, careful choice of the side chains allows hydrophobic surfaces to be produced having easy-to-clean properties. Fluorinated alkoxysilanes such as 1H,1H,2H,2H-perfluorooctyltriethoxysilane (FTS) are used for this. The resulting modified sol-gel layers have water contact angles of approx. 115° and in practice are found to have low dirt uptake and easy-to-clean properties [2]. Other examples are given in Chapter 3.1, including photocatalytic layers produced using the sol-gel process.

Hydrophilic surfaces can also be produced using sol-gel systems. For example, anti-fog layers can be produced. Any water spreads out on these surfaces to form a thin film rather than drops and also means that incident light, for example on a pane of a glass, is not irregularly scattered or reflected. This has been successfully achieved with titanium dioxide (anatase) filled photocatalytic layers, and also with surfactants on the surface. The latter allows the longevity of the anti-fog effect to be significantly improved, and involved incorporating nanoparticles as internal reservoirs of surfactant into an inorganic oxide network (for further details see ref. [1]). These layers have a water contact angle of <15° and are used as anti-fog layers on the inside of car headlights, on spectacles, on all types of visors, and in optical instruments.

Sol-gel based layers have also been described as being reservoirs for other surface functionalities, so enhancing the effects. Ayres et al. (2007) describe, for example, an anti-icing coating based on a titanium alkoxide sol-gel system filled with tripropylene glycol and glycerol as freezing point depressors. The use of freezing point depressors suppresses ice formation on surfaces (see also Chapter 3.2.2.1). The sol-gel system that was developed resulted in slower release of the agents and hence to a prolonged effect [3]. Other examples of the application of sol-gel coatings are described in Chapter 3.6.3.3 and elsewhere, including as a matrix for living microorganisms.

There are an enormous variety of potential applications for coatings based on the sol-gel process. The resulting coating and layer properties are often superior to those of purely organic-based coatings. The disadvantages associated with the inorganic network, such as lack of flexibility/brittleness and the high curing temperatures, can be minimised by incorporating organic components. Here, the optimal balance of components for the sol-gel coating must be determined for each individual application.

2.4.3 Literature

[1] *S. Sepeur*, Nanotechnology: Technical Basics and Applications, Vincentz Network, European Coatings Tech Files, Hanover, 2008

[2] DE 19917367 A1,“ Verfahren zur Herstellung von Überzügen auf Basis flourhaltiger Kondensate, 19.10.2000

[3] *Ayres, J., Simendinger, W.H., Balik, C.M.*, Characterization of titanium alkoxide sol-gel systems designed for anti-icing coatings: II. Mass loss kinetics, J. Coat. Technol. Res., 4 (4), 2007, 473-481

2.5 Polymer brushes

2.5.1 Principles of polymer brushes

The most important surface properties that can be used to create new surface functions are wettability (surface energy), adhesion, switchability, adaptation to ambient lubrication (friction), biocompatibility, and specific microstructure and nanostructure. Recent research on polymer brushes has shown that all these functions can be realised, at least under laboratory conditions, through functional polymer brushes [1]. The term "polymer brush" refers to a polymeric system in which chains of polymer molecules are attached via one or more anchor points to a surface in such a way that the graft density of the polymers is high enough for the surface-attached chains to become crowded and stretched away from the surface [2]. Under laboratory conditions means that the substrates to be coated are, for example, gold surfaces [3], silicon wafers [4], or other well defined surfaces. In contrast to the typically very thin (several nm) self-assembled monolayers (SAMs), which are widely used for adhesion promotion or pretreatment for corrosion protection systems, polymer brushes have more versatile functions and better durability. The main differences between SAMs and polymer brushes with respect to functional coatings are:

- **Thickness**: The thickness of SAMs is typically several nanometres, which makes them mechanically and chemically fragile [5]. Polymer brushes can have thicknesses up to several microns which gives them significantly higher mechanical and chemically resistance.
- Using SAMs, **functional chemical groups** can only be introduced on the surface, whereas polymer brushes can carry functional groups all along the polymer-chain, namely also inside the layer.
- Due to the relatively small size of the molecules, SAMs can introduce a **high density of functional groups** on the surface, but the much longer molecules of polymer brushes can respond to external triggers (e.g. solvents, temperature, pH, etc.) due to the mobility of the polymer chains.
- In general, polymer brushes can be applied on flat surfaces or on the surfaces of particles, and preferably on nanoparticles (see Figure 2.43).

The ability to respond to external triggers such as temperature or changes in pH, or the ability to adapt to environmental conditions, can be achieved through different technical approaches. Figure 2.44 shows two different types of polymer brushes ("black-type" and "grey-type") having different properties that are arranged in nanometre-sized, lateral domains on the surface. Dependent on the external stimulus, the "black-type" brushes are in a stretched conformation, and the "grey-type" brushes are in a compressed conformation. Consequently, the entire surface in this case has the properties of the "black-type" brushes. Of

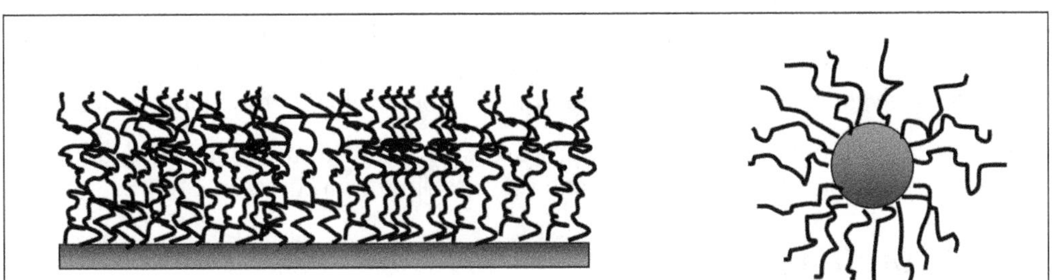

Figure 2.43: Polymer brushes on flat surfaces and particles

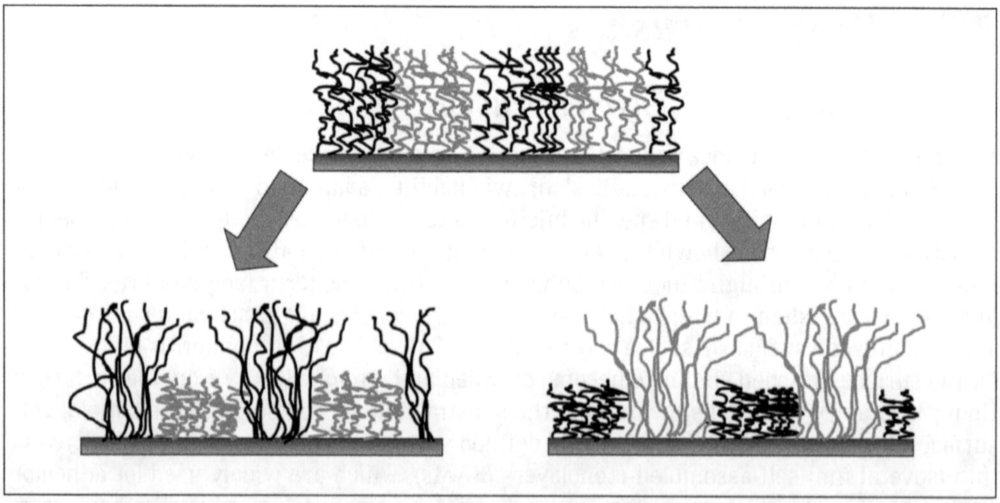

Figure 2.44: Working principle of switchable surface properties based on lateral domains of different polymer brushes

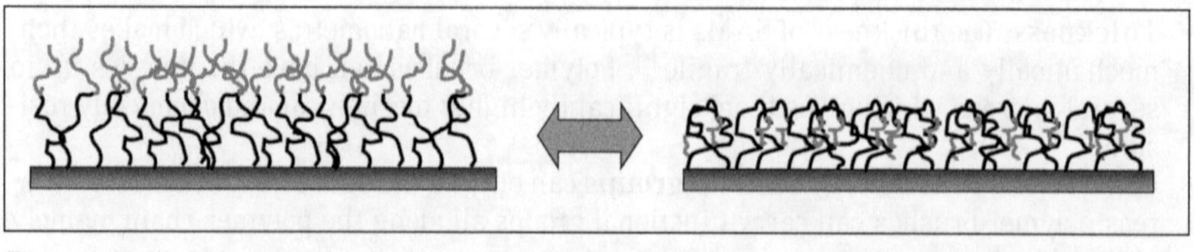

Figure 2.45: Working principle of switchable surface properties using brushes based on block-copolymers

course this process could also work the other way, leading to surface properties dominated by the properties of the grey brushes. Stimuli could, for example, be temperature, humidity, pH, etc. Properties of the brushes with respect to functional surfaces could, for example, be wettability, affinity to special substances, surface charge, etc.

Another option is to use polymer brushes based on block-copolymers. These block-copolymers are built from segments consisting of monomers with different properties. Dependent on the external stimulus, the "black-type" brush segments are orientated towards the surface and dominate the surface properties. Under different conditions, the "grey-type" segments are responsible for the properties of the outer surface (Figure 2.45).

The use of these various polymer brushes (homogeneous brushes, switchable brushes based on lateral domains of different brushes, switchable brushes based on block-copolymers) represents a very versatile "tool-box" for the development of tailor-made functional surfaces. Example applications are given in the following. In addition to the before-described simple principles of polymer brushes, other brush-types such as diblock-copolymers, random copolymer brushes, charged homopolymer brushes, hyperbranched brushes, etc. also exist [6]. Each of these systems is creating new opportunities for developing functional surfaces.

How, though, are these kinds of surfaces prepared? Two methods are established for preparing polymer brushes: The "grafting-to" and the "grafting-from" process.

2.5.2 "Grafting to" process

The basic idea of the "grafting to" process is simple: "Grafting to" means that polymer-chains are attached to the surface from a solution, using certain anchor-groups on the

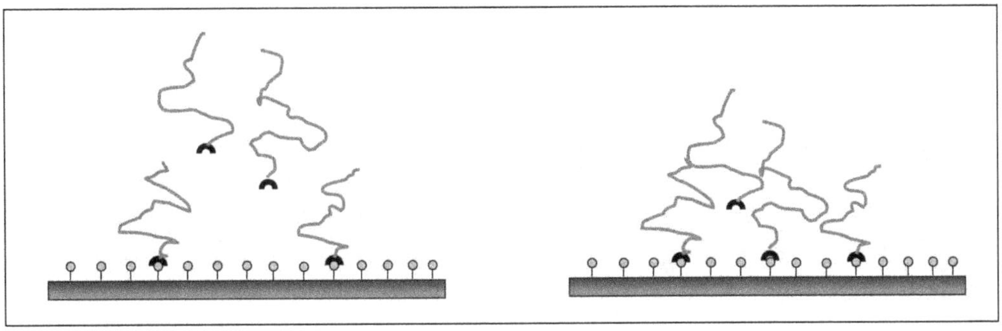

Figure 2.46: Schematic representation of the "grafting to" process

surface and at the end of the polymer chains. This process can be carried out using a simple dipping procedure or also, for example, using a spin-coating method [7]. The anchor groups on the surface can be intrinsic (e.g. SiO_2 groups on silicon wafers, gold atoms, functional organic groups on polymer surfaces) or they can be created with high density using self-assembled monolayers (SAMs) that carry the relevant anchor groups.

Due to the fact that the bulky polymer chains have to reach the surface with their anchor-groups, the surface grafting density is limited. The first chains can easily reach the surface, but with increasing grafting density the reaction is hampered due to steric hindrance (Figure 2.46). The conformation of the polymer chains is, amongst other things, dependent on the grafting density. In other words, there needs to be a certain grafting density to get the polymer chains tethered to the surface as polymer brushes (Figure 2.47).

In order to achieve this kind of high grafting density, constructive interaction between the polymer chains is essential, because a gain in enthalpy from the polymer chain interactions is necessary in order to overcome the entropic effect of the stretching of the chains (which is the same entropic effect as stretching rubber). For most applications the brush conformation is important and in all cases the grafting density has to be optimised for the given application [8].

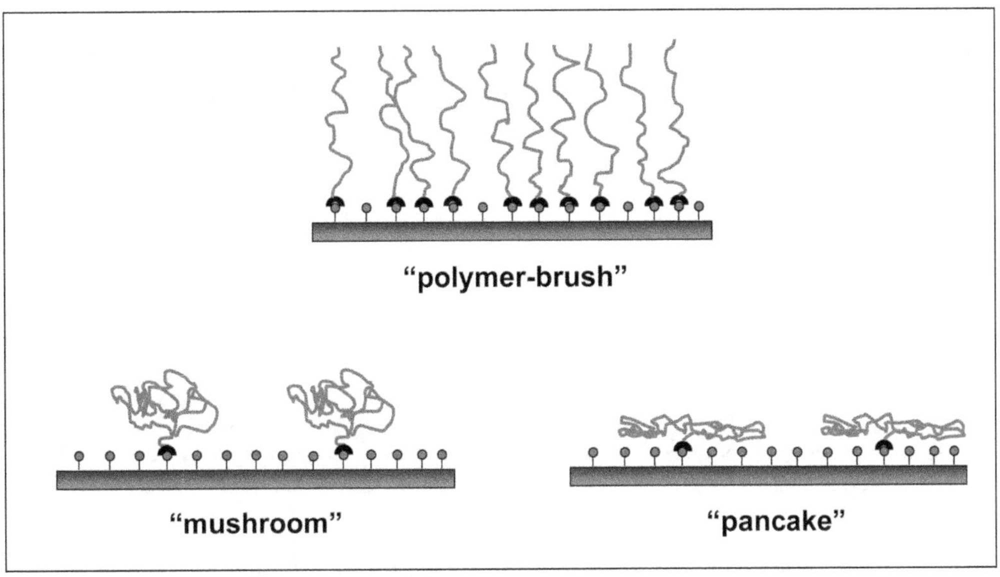

Figure 2.47: Different conformations of tethered polymer chains

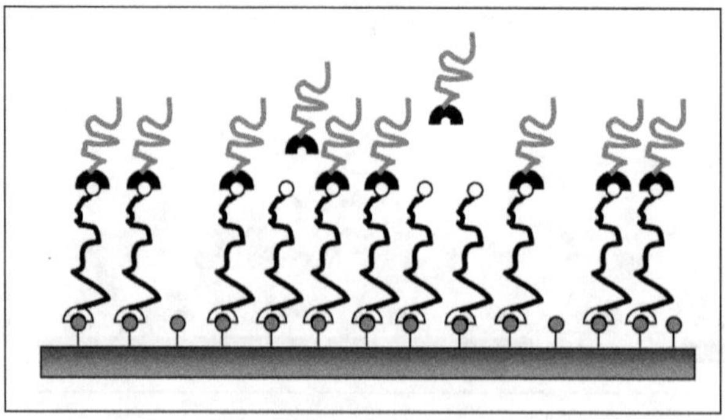

Figure 2.48: Two-step "grafting to" process

The layer thickness that can be achieved for these kinds of polymer brushes is about 10 nm (in contrast to the "grafting from" process where much thicker layers can be prepared (see in the following)).

The process can be carried out as a two-step process or also as a multi-step process resulting in polymer brushes with different distinct segments (Figure 2.48).

Examples of suitable anchor group reactions for "grafting to" processes are:

- Gold-thiol (HS-R)
- Silica (Si-OH) – methoxy silane (e.g. 3-glycidoxypropyl trimethoxysilane-GPS) [9]
- Amine groups (e.g. generated by plasma treatment of polymers) – carboxyl groups of organic acids [10]

A typical application of the "grafting to" process is the functionalisation of surfaces via the immobilisation of proteins (see Chapter 2.6).

The process of surface modification via the "grafting to" method can be carried out comparatively easily, but unfortunately it has some inherent disadvantages:

Due to the fact that the approaching polymer chains have to overcome the barrier formed by the already tethered polymer chains, the surface coverage and the layer thickness are limited, as stated earlier. The limiting factor here is the fact that the bulky polymer chains have to diffuse to the reactive surface. If thicker and denser polymer brushes are desired, then a different process can be applied in which only small monomers have to migrate to the growing polymer brush layer: the "grafting from" process.

2.5.3 "Grafting from" process

This approach uses a layer of surface-immobilised initiator, followed by in situ polymerisation, to generate the polymer brush. Using this technique it is possible to prepare thick polymer films which are covalently anchored to the substrate with a high grafting density [11]. Due to

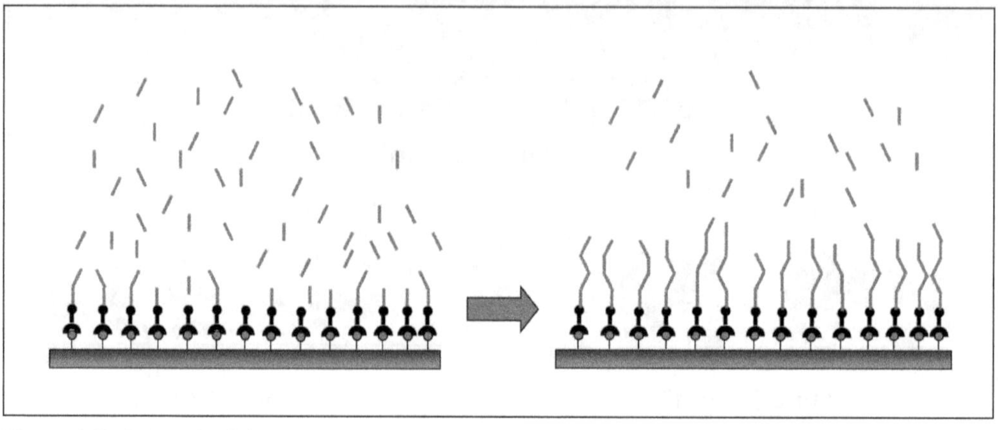

Figure 2.49: Principle of the "grafting from" process; first: surface-immobilised polymerisation initiators, second: monomers

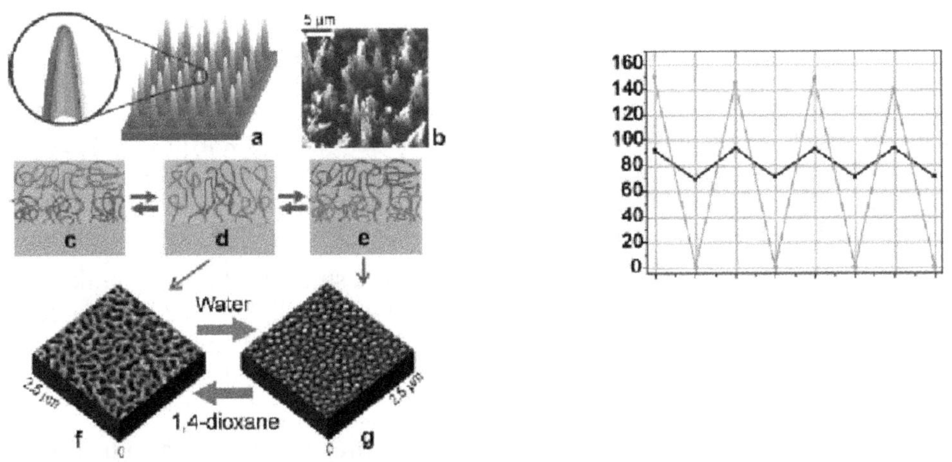

Figure 2.50: Amplification of the switching amplitude by a combination of a microstructured surface and polymer brushes, right: switching of water contact angle (vertical axis), black line: smooth surface, grey line: rough surface [15] *source: Elsevier Ltd., Oxford, UK*

the fact that the monomers here can easily reach the reactive sites of the growing polymer, there is no significant "self-hampering" of the reaction. Consequently, compared to the "grafting to" processes, there is less limitation regarding grafting density and layer thickness.

The process depends on the immobilisation of polymerisation initiators on the surface which is to be functionalised with brushes. In the next step the polymerisation starts directly at the surface-bound initiators and a polymer brush layer forms (Figure 2.49).

Earlier strategies were mostly based on conventional free radical polymerisation (FRP) which was successful in terms of forming grafted films of increased density. The problem was that the control of the chain length and chain length distribution is very poor using this technique.

In order to overcome this problem, living polymerisation techniques including anionic, cationic, ring-opening, ring-opening metathesis, and living radical polymerisation have recently been successfully applied to surface-initiated graft polymerisation [12].

The common feature of these techniques is that they allow controlled growth of the polymer brushes, which leads to a more homogenous chain length and opens up the possibility of preparing functional block-copolymers (see before). The preparation and properties of polymer brushes using this technique have been widely studied by different research groups [13, 14].

2.5.4 Practical application of polymer brushes

Research activities in the field of polymer brushes are highlighting the opportunities which are being opened up by this technology. The following examples serve to illustrate the new surface functions which can be achieved using polymer brushes.

2.5.4.1 Reversible switching between super-hydrophilicity and super-hydrophobicity

Figure 2.44 shows the principle of switching between surface properties dominated by one type of polymer brush (e.g. "black type") and surface properties given by another type of brush (e.g. "grey type") which form lateral domains on the surface. The switching can be triggered by the choice of solvent: If one solvent is a good solvent for the "black type" brush and a bad one for the "grey-type" brush, this leads to a conformation where the "black type" brushes, in contrast to the "grey-type" brushes, are stretched and consequently dominate the surface properties. If the "black type" brushes are hydrophobic and the "grey type" brushes hydrophilic, the described conformation would result in a hydrophobic surface. With another choice of solvent (good for the "grey type" brushes and bad for the "black type"), this would result in a hydrophilic surface. It has been demonstrated that this kind of switching is reversible [15].

If this kind of polymer brush arrangement is applied to a surface with a needle-like microstructure (in this example prepared by plasma-etching of PTFE, Figure 2.50), this results in an amplification of the switching amplitude (see also Chapter 2.1) and leads to reversible switching between super-hydrophilicity (water contact angle 0°) and super-hydrophobicity (water contact angle 150°) [16, 17].

Switching between super-hydophilicity and super-hydrophobicity can also be achieved using temperature as a trigger. Recent research work has shown that special polymer brushes grown by atom transfer radical polymerisation (one of the "grafting from" methods) on a rough silicon substrate have super-hydrophilicity (water contact angle of 0°) at 25 °C, and switch to super-hydrophobicity (water contact angle of 150°) when the temperature reaches 40°C [18].

Surfaces which are able to switch between being hydrophobic and hydrophilic are of interest for a number of technical applications, for example easy-to-clean surfaces, anti-soiling surfaces for clothing, and the surfaces of printing cylinders.

2.5.4.2 Surfaces with controlled behaviour against biological systems

The control of the interactions between biological systems and surfaces is an important issue for several technical applications. For implants, for example, it is vital that the surface of the implant is highly compatible with the surrounding tissue. On the other hand, antibacterial properties are desired for articles to be used for surgery, and for the prevention of marine fouling there is still a lack for environmentally compatible solutions. Several research studies have looked at the use of polymer brushes for such applications. In the following chapters a few examples will be given to illustrate that work.

Titanium is a widely used material for medical applications due to its strength, corrosion resistance, and biocompatibility. Unfortunately, there is often unspecific adhesion of proteins or cells on the oxide surface and this is detrimental for applications in environments which are supposed to be sterile. Studies have been undertaken with polyethylene glycol (PEG) based polymer brushes, prepared with high density and a layer thickness of approximately 100 nm using the "grafting from" approach on titanium surfaces. It was demonstrated that cell-fouling was reduced significantly compared to fouling on unmodified titanium surfaces [19].

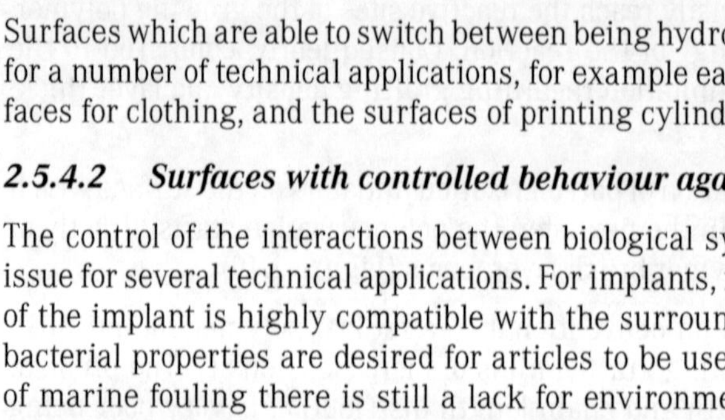

Figure 2.51: Immobilisation of functional peptides on polymer brushes (see also Chapter 2.6)

Another promising approach for using polymer brushes to produce anti-fouling surfaces has been investigated recently. In this case the polymer brushes were used

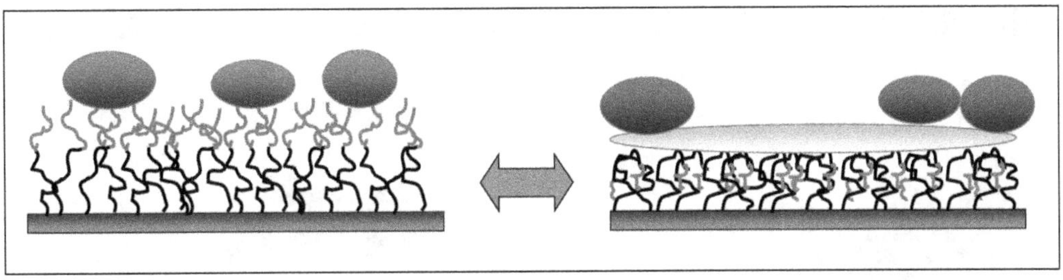

Figure 2.52: Working principle of the removal of dirt from surfaces that are super-hydrophilic and oleophobic

for the immobilisation of natural anti-microbial peptides [20]. To produce these surfaces, polymer brushes were grown on silicon wafers using the "grafting from" approach. Subsequently a natural defence protein ("magainin") was tethered to the free ends of the polymer chains (Figure 2.51). The result was non-adhesive behaviour, delivered by the polymer chain, to reduce cell adhesion, combined with the biocidal effect of the grafted peptide. The effectiveness of such surface coatings against microbes was demonstrated in the study cited above.

Several further applications of polymer brushes in biotechnology are under investigation, mostly in combination with micro and nano patterning techniques, for example molecular recognition, biosensing, protein separation and chromatography, combinatorial chemistry, scaffolds for tissue engineering, and microfluidics and nanofluidics [21].

2.5.4.3 Reversible switching between super-hydrophilic and oleophobic properties

The combination of super-hydrophilic and oleophobic properties is extraordinarily interesting because on such surfaces water spreads out completely whereas any oil (or oily and greasy dirt) has poor wetting behaviour. As a consequence, oil could be removed by water without any detergents or surfactants, because the water would creep under the oil phase and so remove it from the surface (Figure 2.52).

The key question is how to achieve this kind of behaviour. A very promising approach has been reported recently [22]: On a surface bearing OH-groups (a glass slide), an isocyanate-

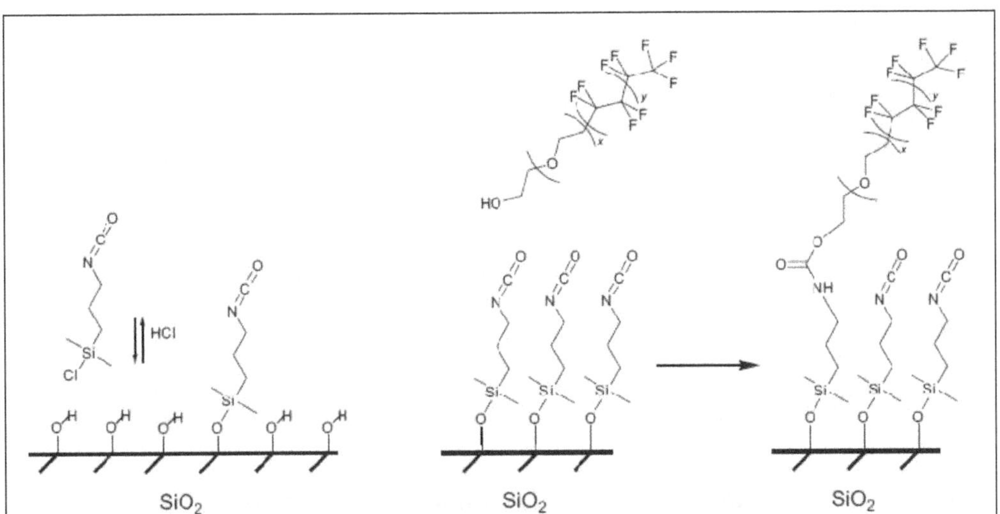

Figure 2.53: Two-step grafting procedure for f-PEG brushes (source of Prof. P. Youngblood, School of Materials Engineering, Purdue University, West Lafayette, IN 47907, USA, with permission of Wiley-VCH)

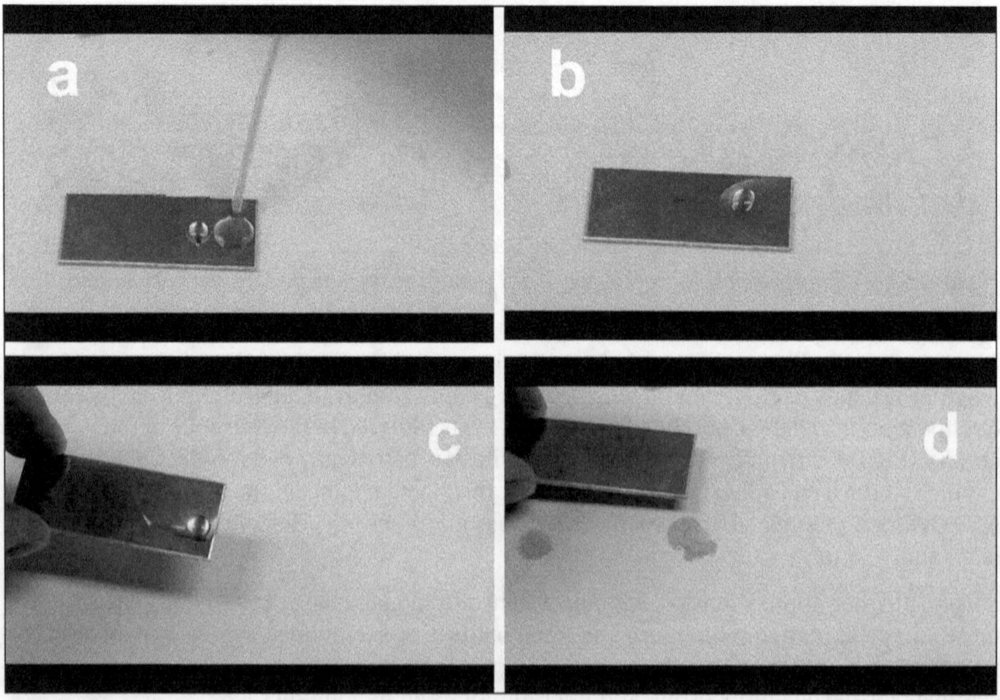

Figure 2.54: Self-cleaning response: a:oil droplets are placed on the surface followed by water (slightly tilted), b: water is creeping under the oil drop, c: with minimal agitation, water displaces the oil, d: upon tilting the sample, the oil floats off and is removed from the sample source: Fraunhofer IFAM

functionalised silane monolayer was formed, followed by a "grafting to" process with polyethylene glycol chains that carry short perfluorinated end caps ("f-PEG") (Figure 2.53).

If a suitable grafting density is achieved, where the density is high enough to give the functionalisation and low enough to give the polymer chains the space for the necessary movement, then the desired super-hydrophilic and oloephobic behaviour can be obtained. If the surface is covered with oil, the oleophobic (and hydrophobic) perfluorinated end caps of the polymer chain are aligned towards the surface, resulting in a relatively high contact angle for oil (approx. 80° for hexadecane). In contact with water, the orientation of the chain segments changes: The hydrophilic PEG segments of the polymer chain form the outer surface, leading to super-hydrophilic properties. Figure 2.54 shows the effect in a laboratory experiment, in which the brushes are not grafted to the surface, but included in the paint formulation.

Surfaces with these kinds of polymer brushes also possess interesting anti-fog properties. The main feature here is again the ability to clean the surface purely with water in the event of contamination with oily substances (e.g. grease from fingerprints), in order to restore the anti-fog effect.

This is a striking example of the many opportunities that exist for functional surfaces created by polymer brushes, but the reality is that the mass industrial production of such surfaces is still a few years away.

2.5.4.4 Outlook for the industrial application of polymer brushes

The practical application of polymer brushes for functionalising surfaces for industrial goods is just at the outset. Potential surface functions have typically been demonstrated to date in laboratory-scale experiments on substrates such as gold surfaces, silicon wafers, and glass slides. The challenge for the coming years is clearly to find ways of transferring these techniques to substrates of industrial relevance, developing industrial processes for the surface treatment, and ensuring that the durability of these surface functions meets the expectations of consumers.

2.5.5 Literature

[1] *Uhlmann, P.; Ionov, L.; Houbenov, N.; Nitschke, M.; Grundke, K.; Motornov, M.; Minko, M.; Stamm, M.*; Surface functionalization by smart coatings: Stimuli-responsive binary polymer brushes, Progress in Organic Coatings, 55, 2006, 168-174

[2] *Advincula, R. C.; Brittain, W.J.; Kenneth C.; Rühe, J. (Eds)*, Polymer Brushes, Wiley-VCH, Weinheim, 2004, Page 15

[3] *Hühnle, A.*, Current Opinion in Colloid & Interface Science, 14, 2009, 157-168

[4] *Baum, M.; Brittain, W. J.*, Synthesis of polymer brushes on silicate substrates via reversible addition fragmentation chain transfer technique, Macromolecules, 35, 2002, 610-615

[5] *Zhou, F; W.T.S. Huck*, Surface grafted polymer brushes as ideal building blocks for "smart" surfaces, Physical Chemistry Chemical Physics, 8, 2006, 3815-3823

[6] *Zhao, B.; Brittain, W.J*; Polymer brushes: surface-immobilized macromolecules, Progress in Polymer Science, 25, 2000, 677-710

[7] *Sirard, S.M.; Gupta, R.R.; Russel, T.P.;Watkins, J.J.; Green, P. F.; Johnston, K.P.*, Structure of End-Grafted Polymer Brushes in Liquid and Supercritical Carbon Dioxide: A Neutron Reflectivity Study, Macromolecules, 36, 2003, 3365-3373

[8] *Brittain, W.J.; Minko, S.*, A structural definition of polymer brushes, Journal of Polymer Science: Part A: Polymer Chemistry, 45, 2007, 3505-3512

[9] *Minko, S.; Patil, S.; Datsyuk, V.; Simon, F.; Eichhorn, K.-J.; Motornov, M.; Usov, D.; Tokarev, I., Stamm, M.*, Synthesis of adaptive polymer brushes via "grafting to" approach from melt, Langmuir, 18,1, 2002, 289-296

[10] *Motornov, M.; Minko, S.; Nitschke, M.; Grundke, K.; Stamm, M.*, Mixed polymer brushes on polyamide substrates, Polymeric Materials: Science & Engineering, 88, 2003, 264-265

[11] *Ayres, N., Cyrus, C. D, Brittain, W.J.*, Stimuli-responsive surfaces using polyampholyte polymer brushes prepared via atom transfer radical polymerization, Langmuir, 23, 2007, 3744-3749

[12] *Tsuji, Y.; Ohno, K.; Yamamoto, S.; Goto, A.; Fukuda, T.*, Structure and properties of high-density polymer brushes prepared by surface-initiated, living radical polymerization, Advamced Polymer Science, 197, 2006, 1-45

[13] *Baum, M.; Brittain, W.J.*, Synthesis of polymer brushes on silicate substrates via reversible addition fragmentation chain transfer technique, Macromolecules, 35, 2002, 610-615

[14] *Edmondson, S.; Osborne, V.L.; Huck, W.T.S*, Polymer brushes via surface-initiated polymerizations, Chem. Soc. Rev., 33, 2004, 14

[15] *Uhlmann, P.; Houbenov, N.; Ionov, L.; Motornov, M.; Minko, S.; Stamm, M.*, Oberflächen passen sich an - bürstenartige Polymermoleküle an Oberflächen mit schaltbaren Eigenschaften, Wissenschaftliche Zeitschrift der Technischen Universität Dresden, 56, 2007, 47-52

[16] *Minko, M.; Müller, M.; Motornov, M.; Nitschke, M.; Grundke, K.; Stamm, M.*, Two-level structured self-adaptive surfaces with reversibly tunable properties, J. Am. Chem. Soc. 125, 13, 2003, 3896-3900

[17] *Nitsche, M.; Minko, S.; Motornov, M.; Grundke, K.; Stamm, M.* Oberflächen mit schaltbarer Ultrahydrophobie und Verfahren zur Realisierung, DE10206481, EP1350575

[18] *Xia, F; Zhu, Y; Feng, L; Jiang, L*, Smart responsive surfaces switching reversibly between super-hydrophobicity and super-hydrophilicity, Soft Matter, 5, 2009, 275-281

[19] *Fan, X; Lin, L; Messersmith, P.B.*, Cell fouling resistance of polymer-brushes grafted from Ti substrates by surface-initiated polymerization: Effect of ethylene glycol side chain length, Biomacromolecules, 7, 2006, 2443-24482.7

[20] *Glinel, K.; Jonas, A.M., Jouenne, T.; Leprince, J.; Galas, L.; Huck, W.T.S.*, Antibacterial and antifouling polymer brushes incorporating antimicrobial peptide, Bioconjugate Chem., 20, 2009, 71-77

[21] *Senaratne, W.; Andruzzi, L.; Ober, C.K.*, Self-assembled monolayers and polymer brushes in biotechnology: Current applications and future Perspectives, Biomacromolecules, 6, 2005, 2427-2448

[22] *Howarter, J.A.; Youngblood, J.P.*, Self-cleaning and anti-fog surfaces via stimuli-responsive polymer brushes, Adv. Mater., 19, 2007, 3838-3843

2.6 Biologically inspired coatings

The toolbox of methods for functionalising surfaces is extremely comprehensive and not all approaches can be covered in this book. One key area which was already touched upon in previous chapters does however deserve further mention here. This concerns the modification of surfaces following the example of nature, and includes the modification of both the structure and chemistry of surfaces. As the structuring of surfaces has already been described in detail elsewhere (Chapter 2.1), the current section will focus on biochemical/chemical modification via the immobilisation of active substances on surfaces.

Bio-mimetic coatings, in the sense meant here, have additional functions due to the attachment of biologically inspired molecules to their surfaces (for further information about the functions see, for example, Chapter 3.6). At this juncture, we will first of all consider the production and recovery of the relevant molecules and their immobilisation on surfaces. The simplest form of attachment, namely the physical adsorption of molecules on surfaces (for example via dip coating or spray application), is not mentioned further here because in general this solely involves non-covalent attachment of the molecules. As such, important preconditions for effective bio-mimetic coatings are not met, because there is no oriented surface attachment of the molecules (and hence the active centres of the biomolecules can be blocked). In addition, only covalent bonding guarantees a stable and durable coating. The available methods include the entrapment or attachment of molecules (in insoluble beads or microspheres, nanoparticles, and capsules) as well as direct chemical grafting on the surface. These methods will be individually described here.

2.6.1 Recovery of biofunctional molecules or production of synthetic analogs

There are many methods for producing and recovering biofunctional molecules. These substances belong to different classes of materials, and include antibiotics such as penicillin, alkaloids such as polyalkylpyridinium salts (anti-fouling agents), and proteins and peptides. As the chemical synthesis and biotechnological production of these materials are fields of science in their own right, we will only briefly describe the methods here and will focus on examples of relevance for this book.

The most obvious method is to directly recover the active substances from organisms that synthesise them. This is carried out, for example, with marine red seaweed *Chondrus crispus*, which is used amongst other things as a source of carrageenan (a thickening agent) in the food industry and is hence processed industrially in large quantities. Substances in this species also have anti-fouling effects, meaning that extracts from the dried starting material have been incorporated into anti-fouling coating systems [1]. The requirements on the purity of the active substances here was relatively low, allowing this simple extraction method to be used.

For other substances produced by organisms, direct recovery is either not possible or is very uneconomical. An example of this concerns the adhesive proteins produced by the blue mussel *Mytilus edulis*. These proteins possess extraordinary adhesive properties which, if transferred to technical applications, would allow significant improvement over currently used systems. The recovery of this raw material is however very complex and time-consuming: For 1 g one requires about 10,000 animals [2]. Methods must thus be developed which allow the large-scale production of the active components, namely the synthetic production of these molecules.

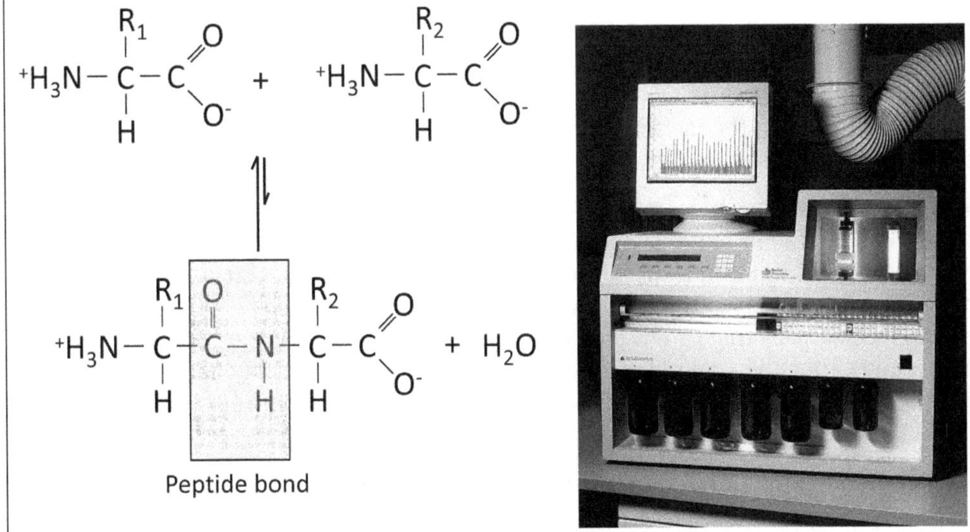

Figure 2.55 left: Base formula of amino acids and their linking to form peptides or proteins via peptide bonds; right: Solid phase peptide synthesiser for manufacturing functional peptides source: Fraunhofer IFAM

First of all there are biotechnological methods which in general use cells or bacteria for producing substances. These are cultivated in bioreactors under optimal conditions and produce the desired substances on a large scale. Examples are the production of antibiotics (penicillin using Penicillium fungi) and bioethanol production from cereal starch (alcoholic fermentation). It is also possible to use genetically modified organisms or cells containing specific gene sequences which are responsible for the synthesis of the desired molecules. For example, enzymes and insulin can be produced in this way (for further background reading see relevant teaching books such as Berg et al. (2002) [3]).

Another method which is used in particular for the synthesis of peptide sequences is solid phase peptide synthesis (SPPS). Peptides are biological molecules composed of amino acids which are linked to each other via peptide bonds (see Figure 2.55 left). Up to about 50 amino acids can be joined together in this way in peptides. If there are more than 50 amino acids then the resulting molecules are generally called proteins. The amino acid sequence (with their characteristic side chains) largely determines the spatial arrangement of the atoms in the molecule and hence its activity. In SPPS a peptide is built up step by step: one amino acid at a time. Here, the individual synthesis steps occur on fixed amino acid chains, meaning that the subsequent isolation and purification steps of the intermediate and end products is easier. This method is used in a standard way in biochemical laboratories (Figure 2.55 right shows a solid phase synthesiser) and also in industry for the manufacture of peptide pharmaceuticals on a tonnage scale [4]. Peptides of relevance for the functionalisation of surfaces which can be produced using this method include adhesive peptides and active sequences of anti-freeze proteins.

2.6.2 Chemical attachment of biofunctional molecules to surfaces

Many methods can be used to attach bioactive compounds to surfaces. A fundamental distinction is first of all made here between non-covalent and covalent attachment mechanisms. Non-covalent bonds arise, for example, when molecules are adsorbed via electrostatic interactions. These non-specific and relatively weak bonds are not considered in more detail here as they produce considerably less stable layers than covalently bonded systems. Indeed, covalent attachment allows strong and stable bonds to form between the molecules and the surface, assuming the selected reaction partners are compatible. It must also be ensured the activity of the molecules is not impaired by the immobilisation, so preserving the desired surface functionalities.

Various example methods for immobilising molecules are described below. Topics covered include the preparation of the substrate surfaces (activation and generation of functional groups), the use of linker and spacer molecules, and the final attachment of the bioactive compounds. The examples that are given will give insight into some of the fundamental chemical mechanisms that can be employed. Some of the methods are, however, extremely complex and cannot be comprehensively covered in this book. A detailed introduction to this topic has been given by, amongst others, Goddard und Hotchkiss (2007) [5] and Wong et al. (2009) [6].

2.6.2.1 Generation of reactive functional groups: surface activation and linker molecules

In order to allow covalent attachment of bioactive compounds to surfaces, compatible reactive groups must be present on the reaction partners. Categories of functional groups of importance for this attachment include thiol groups, aldehyde groups, carboxylic acid groups, hydroxyl groups, and primary amine groups.

Diverse substrate materials can be used for the chemical attachment of biomolecules. Research work is often undertaken on noble metals such as gold (bond type: thiol, HS-R), glass (Si-OH, e.g. in connection with self-assembled monolayers (SAMs) based on silane), and silicon wafers (reactive SiO_2 groups). These materials possess well-defined surfaces and usually allow stable bonding of the molecules to the substrate. Polymer materials are also the focus of many practical applications. These surfaces are naturally inert and so must be activated using pretreatment techniques. Wet chemical methods can be used for this purpose, as highlighted by the following examples [5]:

- Carboxylic acid groups can be generated via base or acid hydrolysis of PMMA using concentrated sodium hydroxide or sulfuric acid.
- Primary amine groups can be generated on, for example, poly(urethane) surfaces using diamines such as hydrazine hydrate and ethylenediamine.

Plasma treatments can also be used to generate reactive groups on polymer surfaces, for example oxygen containing groups or amine groups.

A further promising way of generating functional groups is to use coating formulations which, due to deliberate over-crosslinking or under-crosslinking, generate reactive surface groups on the cured coating surfaces for subsequent immobilisation steps. For example, when formulating a 2-C PUR system (based on polyacrylate (OH groups) and polyisocyanate), careful selection of the mixing ratios of the components gives an excess of either hydroxyl or isocyanate groups. These are subsequently available as reactive groups. Here it must be determined whether the reactivity of the end groups suffices for the subsequent attachment step [7].

A further method for creating reactive surface groups is silanisation using organosilanes. These molecules generally consist of a head group having affinity for inorganic substrate surfaces, so allowing an ordered, unimolecular layer to form (also called a self-assembled monolayer (SAM)). Substrates that can be used here include glass and silicon (via siloxy

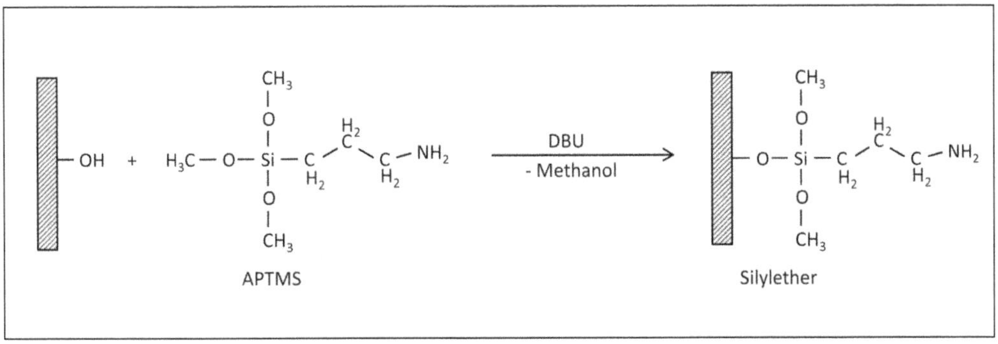

Figure 2.56: Silanisation of a hydroxy-functionalised surface with 3-aminopropyltrimethoxysilane (APTMS) in the presence of diazabicycloundecene (DBU) [7]

linkages), and gold, copper, and silver (via thiol linkages) [5]. In addition, the aforementioned (reactive) surfaces of coatings can also be modified via silanisation. Figure 2.56 shows one possible reaction that is suitable for the silanisation of (coating) surfaces. The resulting improvement of the stability to hydrolysis is the topic of much ongoing research [7].

Besides the head groups responsible for the reaction with the substrate surface, organosilanes also have terminal end groups which, if of suitable chain length, can also serve as "links" between bioactive compounds and the substrate surface. Poly(ethylene glycol) (PEG) compounds are also often used for this purpose and this topic will be returned to later.

Besides being a means of creating functional groups, silanisation is also an example of the use of linker molecules (or spacer molecules) for attaching bioactive molecules. These are very important for many biologically-based functional coatings because they can beneficially influence the activity of the bioactive compounds. The following factors play a role here [5]:

- Reduction of the steric hindrance of the bioactive molecules due to increased flexibility, and hence improved bioactivity
- Shielding of the molecules, so suppressing denaturing reactions triggered by interactions with the substrate surface

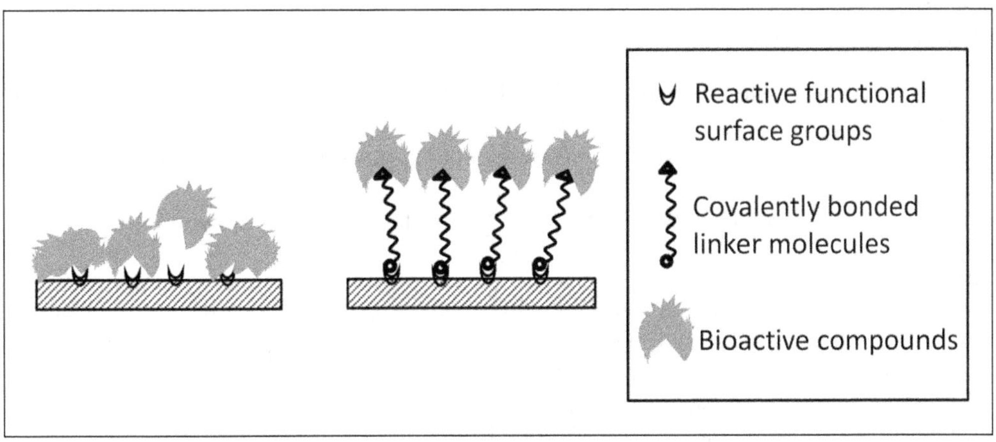

Figure 2.57: Schematic representation of the surface immobilisation of bioactive compounds. Left: Without using linker molecules (steric hindrance, poor flexibility, denaturation processes). Right: Use of linker molecules to improve the bioactivity

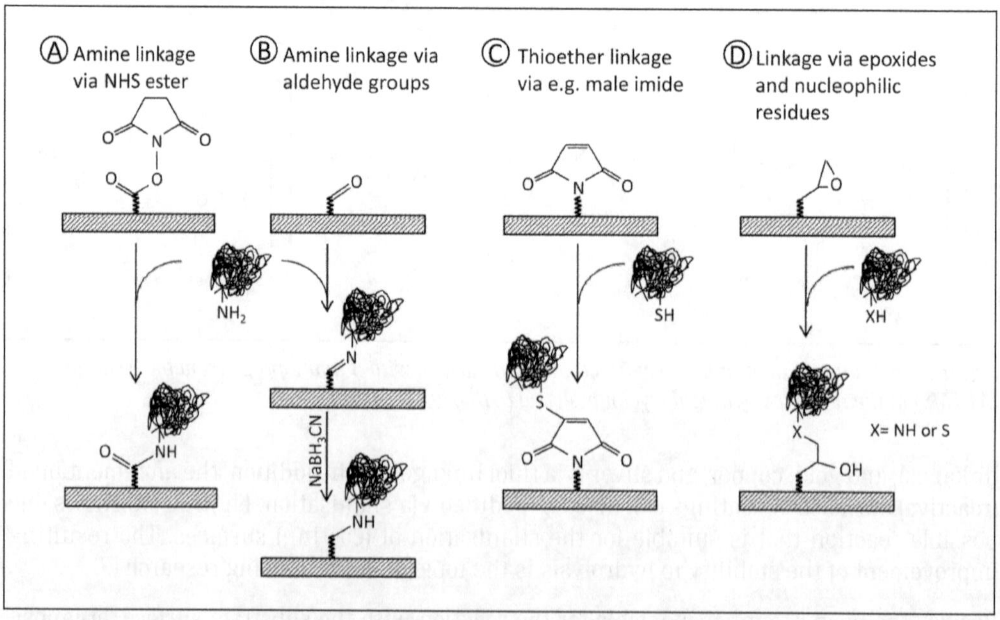

Figure 2.58: Schematic representation of some possible reactions for immobilising proteins [6]

- Increase in the number and accessibility of reactive surface groups and hence increase in the loading density of the bioactive compounds (in particular use of dendrimers)

Poly(ethylene glycol) (PEG) based compounds are often used as linker molecules, and these are also often described in connection with polymer brushes. Noteworthy here is that surface functionalities are already created via these molecules. For example, switchable surfaces can be produced which respond to external triggers, e.g. changeover from hydrophilic to hydrophobic properties (see Chapter 2.5). A further functionality (described in Chapter 3.5) concerns anti-fouling properties which are based on the hydrophilic character of the PEG chains.

2.6.2.2 Attachment of bioactive compounds (immobilisation)

Once the reactive functional groups are available on the substrate surface, compatible reaction partners need to be found (bioactive compounds). Further factors need to be considered when selecting reaction partners:

- There must be no steric hindrance of the bioactive molecule as a result of the covalent attachment
- Conformation changes, caused by the covalent bonds, must be avoided in order to prevent activity loss
- Directional attachment of the molecules at the surface allows optimal alignment of the active sites

Various strategies can be employed for the directional, covalent immobilisation of bioactive compounds. Some of these methods are described below, with special attention being put on the attachment of biologically-derived proteins and peptides. Detailed explanation of the reaction conditions and molecule properties can however not be given here but can be found elsewhere: see, for example, Wong et al. 2009 [6].

The structural properties and hence functionality of proteins and peptides are defined by the amino acid sequence. In total 20 amino acids are found in proteins and peptides, each with characteristic side chains which determine the steric structure of the molecules and hence their activity. The side chains of these amino acids can be used to immobilise the proteins to surfaces. Reaction partners that have been described [6] include the following:

- Side chains of lysine contain amine groups which, if on the surface of the proteins, can react with esters (such as N-hydroxysuccinimide (NHS) in Figure 2.58 A) and form stable compounds. Reaction can also occur via aldehyde groups present on the substrate surface, with the aid of for example sodium cyanoborohydride, in order to reduce the resulting imine to a stable secondary amine linkage (Figure 2.58 B).
- Thiol groups in cysteine residues can be used to form stable thioether bonds with the substrate surface (Figure 2.58 C).
- Proteins with accessible nucleophilic residues (such as lysine and cysteine as already mentioned) can also be attached to epoxide-functionalised materials (Figure 2.58 D).

Other methods which allow covalent, site-specific immobilisation of proteins and have no effect on the electrostatic charges (of carboxyl or amino groups) include Staudinger ligation. The advantage here is that due to preservation of the electrostatic charges in the molecule there is no effect on the protein folding and on the segregation of hydrophobic and hydrophilic domains – and hence the protein stability and activity can be significantly increased. The first step of Staudinger ligation involves labelling the protein of interest with the relevant moiety (e.g. the azide). The azide group subsequently reacts with phosphine to form an iminophosphorane. This then reacts with electrophiles to complete the actual immobilisation reaction. Reaction with esters often results in the formation of stable amide bonds [6].

Other reaction mechanisms for immobilising molecules are based, amongst other things, on "click" chemistry and Diels-Alder cycloaddition. Detailed information on these and other methods can be found elsewhere (Wong et al. 2009 [6]). These complex reaction mechanisms are however only of secondary importance for the manufacture of functional surfaces of relevance for this book.

2.6.2.3 Adhesion promoters mimicking nature

Another way of immobilising molecules is to use natural adhesive proteins derived from DOPA (dihydroxyphenylalanine). DOPA has been identified as a key amino acid in the adhesive cement synthesised by the blue mussel *Mytilus edulis*. This adhesive cement consists of several polyphenolic adhesive proteins, which are responsible for the very effective bonding of these mussels to surfaces [8]. Utilisation of this principle might be beneficial, for example, when developing biocompatible adhesives for medical applications. In the area of functional coatings, these systems could possibly be used to improve the adhesion between the substrate and functional molecules and hence prolong the layer stability and activity.

The reasons for the excellent adhesive and cohesive properties of DOPA are the multitude of possible chemical reactions of DOPA: For example, the catechol moiety can form extremely stable metal complexes, and the hydroxyl groups are hydrogen bond donors and acceptors [8] (see also Figure 3.69 in the section on "Biomimetic surfaces"). One example of the use of DOPA taken from the literature is attachment to PEG chains. As already described, PEG chains are often used for developing new functional coatings. Examples include anti-fouling properties and anti-icing coatings, as well as the switchable surfaces described in Chapter 2.5 which can take on both oleophobic and hydrophilic properties. In addition, PEG chains can be employed to immobilise proteins and act here as linker molecules. Up until now the attachment of PEG chains has been demonstrated on a laboratory scale on surfaces such as gold, silicon wafers, and glass slides. The use of PEG conjugated to DOPA polymers allows stable bonds to be formed with Ti-oxide layers [9]. Titanium is widely used in the medical area, so this represents the first step towards practical applications. A further example

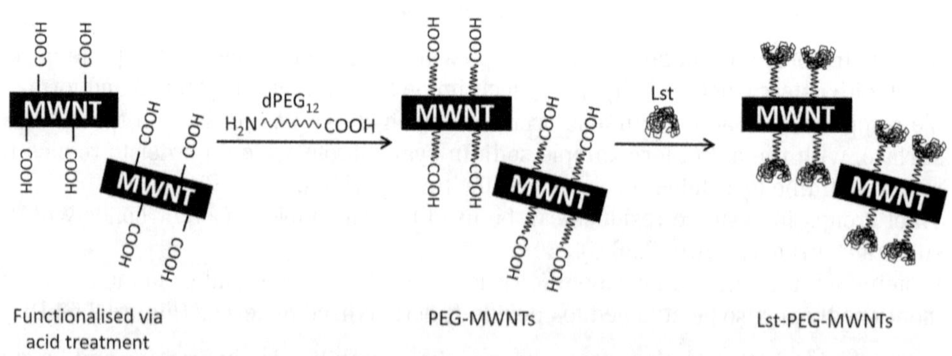

Figure 2.59: Simplified schematic representation of the attachment of the cell lytic enzyme lysostaphin (Lst) to multi-walled carbon nanotubes (MWNT) using PEG linker molecules [14]

of the use of DOPA as an adhesion promoter is the immobilisation of anti-microbial peptoids, as described by Statz et al. 2008 [10]. Chapter 3.6 will cover this in more detail.

2.6.2.4 Example applications

Biologically inspired coatings have a large number of potential applications. The above-mentioned methods are being used and considered, for example, for the manufacture of protein microarrays, biosensors, and continuous flow reactor systems [6]. Protein microarrays can be used, for example, for bioanalytical assays such as mutation analysis, drug discovery, and diagnostics. Enzyme reactors systems are of interest for the food and pharmaceutical industries, because immobilised enzymes allow continuous generation of enzymatic products and the reuse of the enzymes. Active packaging technologies in the food industry are also of interest in order to maintain or improve product quality [5]. This last example involves surface functionalities which have been centre stage in this book, namely coating functionalities which suppress colonisation by microorganisms such as bacteria (anti-fouling and anti-microbial surfaces). These are of great interest for a wide spectrum of applications ranging from the medical area to industrial and marine uses.

The development of anti-microbial coatings based on biologically inspired materials has as its objective the immobilisation of naturally occurring substances on surfaces with preservation of their functionality. For example, cationic anti-microbial peptides (CAPs) have been covalently bonded to surfaces and the effect of this immobilisation on their activity has been studied. The CAPs were manufactured using solid phase peptide synthesis (SPPS) as described in Chapter 2.6.1, and resin beads modified with PEG chains were used as a model solid surface. The covalent attachment occurred at the C-terminus, at the N-terminus and at the side chains, respectively. The subsequent microbial tests gave the following results [11]:

- The activity spectrum of the CAPs towards bacteria was preserved on immobilisation, but the anti-microbial activity was lowered by about 2 orders of magnitude
- The anti-microbial activity is very dependent on the length and nature of the selected spacer molecules
- The cationic charge plays an important role for determining the peptide activity
- The flexibility of the immobilised CAPs, the surface density, and the position of immobilisation are other factors that have effects

$$\text{Starch} \xrightarrow[\text{GA}]{H_2O} \text{D-glucose} \xrightarrow[\text{HOX; } H_2O]{O_2} \text{Glucono-}\delta\text{-lactone} + H_2O_2$$

Figure 2.60: Anti-fouling functionality due to hydrogen peroxide formation via enzyme-catalysed reactions: Starch is first of all converted by glycoamylase (GA) into D-glucose, which in turn is converted to glucono-δ-lactone and hydrogen peroxide in a reaction catalysed by hexose oxidase (HOX) [16]

This acquired knowledge can be transferred from the model solid surface (resin beads) to anti-microbial coatings. The literature contains various examples of incorporating anti-microbial peptides into coatings. For example, the natural defense protein "magainin" was attached to PEG-based polymer brushes linked to silicon wafers. The advantage here, besides the non-adhesive behaviour of the PEG chains, is that these PEG chains also make reactive hydroxyl groups available which allows attachment of anti-microbial peptides [12] (see also Chapter 2.5). Other examples of biologically inspired anti-microbial coatings are given, by amongst others, Chevalier et al. (2010) [13]. The topics of anti-microbial coatings and anti-fouling coatings will be returned to in Chapter 3.5 and 3.6.

Anti-icing surface functionality is a further surface function that will be described in detail in this book (Chapter 3.2). Ongoing studies are investigating whether the immobilisation of biologically inspired anti-freeze proteins affects their bioactivity on surfaces. Further insight into this is given in Chapter 3.6.3.2.

2.6.3 Attachment of biofunctional molecules to particles

Special support materials in the form of particles can also be used to bind biofunctional molecules and then these can be incorporated into coating materials. The particles are usually in the nanoscale range, as already described in detail in Chapter 2.3. The advantages of using support materials are numerous. For example, the molecular structures of the active substances can be stabilised, the anchoring of the molecules can prevent leaching, and negative effects caused by coating components or external factors can be minimised. The use of nanoscale particles also allows high loading of the active materials due to the high surface area-to-volume ratio of the nanoparticles. Examples are described below of the use of fibre-like nanoparticles (CNTs = carbon nanotubes, see also Chapter 2.3.2.1) and spherical particles in biologically modified coatings.

In the first example, the anti-microbial agent lysostaphin (Lst; a cell lytic enzyme) is covalently bonded to carbon nanotubes via the procedure shown in Figure 2.59. Here MWNTs (multi-walled carbon nanotubes) were used, which also contained poly(ethylene glycol) chains (dPEG12). These act as linker molecules in order to be able to guarantee the flexibility and hence effectiveness of the enzyme [14]. The use of PEG chains has already been described in detail elsewhere (Chapter 2.5 and 2.6.2).

The use of these enzyme-nanotube conjugates in latex films gave effective anti-microbial films, whose effect was proven to be not due to leaching effects. This considerably improved the service life compared to films with free Lst [14].

Similar promising results were achieved with poly(methyl methacrylate) (PMMA) based films containing anti-microbial enzymes attached to carbon nanotubes. In this case SWNTs

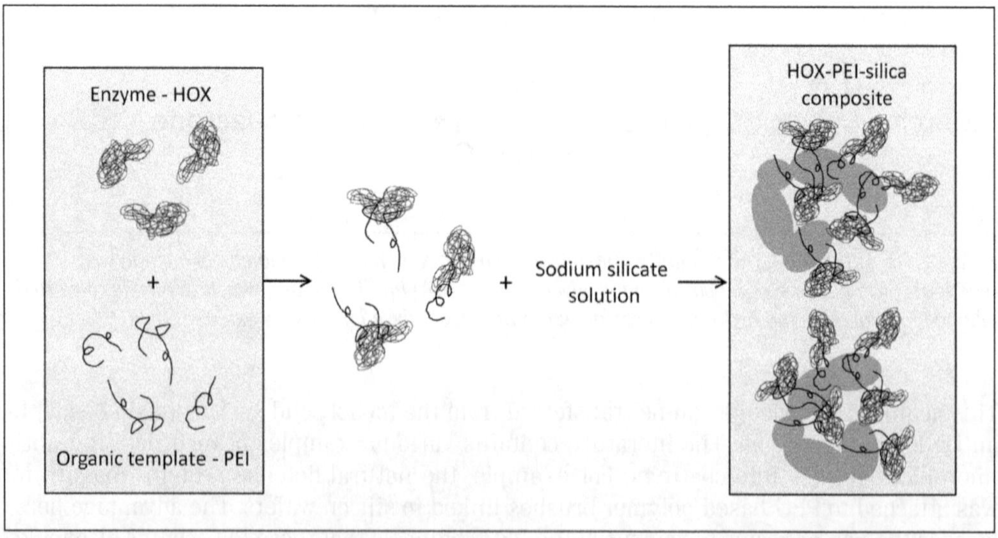

Figure 2.61: Schematic representation of the formation of the enzyme/PEI/silica bio-composites[16]

(single-walled carbon nanotubes) were used, which hindered the leaching of the enzyme and hence led to significantly prolonged effectiveness of the coating. It was also demonstrated that the molecular structure was stabilised by attachment of the enzyme to the particles, meaning that the enzyme activity was also maintained under what would normally be considered unfavourable conditions (e.g. temperatures up to 70 °C, salt solution 3 % (w/v)) [15].

A third example of the development of biologically inspired coatings also concerns an enzyme-containing anti-fouling coating. This is a coating which releases hydrogen peroxide due to the enzyme-catalysed reactions shown in Figure 2.60. The hydrogen peroxide has a toxic effect on the fouling organisms (for further details see Chapter 3.5.2.2).

Studies have also shown that the anchoring of the enzyme HOX via bio-mimetic precipitation leads to the coating having significantly prolonged anti-fouling activity. The following procedure was used here: A HOX suspension was mixed with polyethylene imine (PEI) so that polycation templates formed. Subsequent addition of a sodium silicate solution resulted in the formation of initially small primary particles of enzyme/PEI/silica bio-composite. These ultimately coalesced to form water-insoluble amorphous agglomerates which precipitate out of the aqueous solution. Particle sizes were in the range of 0.7 µm to 17 µm (95 % of the colloidal volume). This so-called silica co-precipitation is schematically shown in Figure 2.61 and was derived from the polycation-templated silicification of diatom and sponge exoskeletons. The HOX-PEI-silica bio-composite was dried and incorporated in powder form with starch and glycoamylase (GA) into commercial coating systems [16].

The examples of molecule attachment to particles that have been described here relate to the development of anti-microbial and anti-fouling coatings. The use of particles is however not limited to these functions, and indeed this approach is also feasible for other functional molecules, such as the anti-freeze proteins described in Chapter 3.2.

These areas are presently the topic of intense R&D work. Current and future developments will determine to what extent and for what specific applications it is possible to use particles with attached biologically inspired molecules.

2.6.4 Literature

[1] *Chambers, L.D.*, The Development of a marine antifouling system using environmentally acceptable and naturally occurring products, Doctoral Thesis, University of Southampton, School of Engineering Sciences, 2008

[2] *Silverman H.G., Roberto, F.F.*, Understanding Marine Mussel Adhesion, Marine Biotechnology, Vol 9, 2007, 661-681

[3] *Berg, J.M., Tymoczok, J.L., Stryer, L.*, Biochemistry, 5th edition, 2002, W.H. Freeman and Company, New York

[4] *Verlander, M.*, Industrial Applications of Solid-Phase Peptide Synthesis - A Status Report, International Journal of Peptide Research and Therapeutics, Vol 13, 2007, 75-82

[5] *Goddard, J.M., Hotchkiss, J.H.*, Polymer surface modification for the attachment of bioactive compounds, Prog. Polym. Sci. 32, 2007, 698-725

[6] *Wong, L.S., Khan, F., Micklefield, J.*, Selective Covalent Protein Immobilization: Strategies and Applications, Chem. Rev. 109, 2009, 4025-4053

[7] *Kreider, A.*, Entwicklung von Immobilisierungsmethoden zur Anknüpfung von Antifreeze-Proteinen an Lackoberflächen, Fraunhofer IFAM Diplomarbeit, 2009, unpublished

[8] *Grunwald, I., Rischka, K., Kast, S.M., Scheibel, T., Bargel, H.*, Mimicking biopolymers on a molecular scale: nano(bio)technology based on engineered proteins, Phil. Trans. R. Soc. A 367, 2009, 1727-1747

[9] *Dalsin, J.L., Lin, L., Tosatti, S., Vörös, J., Textor, M., Messersmith, P.B.*, Protein Resistance of Titanium Oxide Surfaces Modified by Biologically Inspired mPEG-DOPA, Langmuir, 21, 2005, 640-646

[10] *Statz, A.R., Meagher, R.J., Barron, A.E., Messersmith, P.B.*, New Peptidomimetic Polymers for Antifouling Surfaces, J. Am. Chem. Soc., 127, 2005, 7972-7973

[11] *Bagheri, M., Beyermann, M., Dathe, M.*, Immobilization Reduces the Activity of Surface-Bound Cationic Antimicrobial Peptides with No Influence upon the Activity Spectrum, Antimicrobial Agents and Chemotherapy, 53 (3), 2009, 1132-1141

[12] *Glinel, K.; Jonas, A.M., Jouenne, T.; Leprince, J.; Galas, L.; Huck, W.T.S.*, Antibacterial and antifouling polymer brushes incorporating antimicrobial peptide, Bioconjugate Chem., 20, 2009, 71-77

[13] *Chevalier, S., Cuestas-Ayllon, C., Grazu, V., Luna, M., Feracci, H., de la Fuente, J.M.*, Creating Biomimetic Surfaces through Covalent and Oriented Binding of Proteins, Langmuir, 26 (18), 2010, 14707-14715

[14] *Pangule, R.C., Brooks, S.J., Dinu, C.Z., Bale, S.S., Salmon, S.L., Zhu, G., Metzger, D.W., Kane, R.S., Dordick, J.S.*, Antistaphylococcal Nanocomposite Films Based on Enzyme-Nanotube Conjugates, ACSNano, Vol.4 (7), 2010, 3993-4000

[15] *Asuri, P., Karajanagi, S.S., Kane, R.S., Dordick, J.S.*, Polymer-Nanotube-Enzyme Composites as Active Antifouling Films, Small, 3 (1), 2007, 50-53

[16] *Kristensen, J.B., Meyer, R.L. Poulsen, C.H., Kragh, K.M., Besenbacher, F., Laursen, B.S.*, Biomimetic silica encapsulation of enzymes for replacement of biocides in antifouling coatings, Green Chem. 12, 2010, 387-394

3 Examples for functional coatings

3.1 Self-cleaning/dirt-repelling surfaces

A comprehensive survey of German industry in 2006 concerning research needs in the area of surface technology identified nine so-called flagship topics [1]. These are the topics that have the biggest influence on the competitiveness of the companies who took part in the survey. The companies that were surveyed came from a broad range of sectors of industry. One topic is clearly foremost in all these sectors, namely the desire for self-cleaning surfaces.

It is no wonder then that much R&D work has focused on this area in recent years, with the approaches being pursued being as diverse as the requirements of the various industrial sectors.

Fundamentally, there are two classical approaches for developing self-cleaning or dirt-repelling surfaces:

- Hydrophobic (= water-repelling) or superhydrophobic surfaces (*lotus effect, easy-to-clean* surfaces)
- Photocatalytic surfaces

Both approaches have already been described in detail in other books in this series [2]. For this reason, only a brief summary of the classical principles and applications are given here and the main emphasis will be put on more recent research work.

3.1.1 Hydrophobic surfaces, superhydrophobic surfaces, lotus effect

Hydrophobic coatings, having a water contact angle of between 90° and approx. 140°, and superhydrophobic coatings having a water contact angle of > 140° (see Chapter 2.1.2), typically have low dirt uptake and low dirt adhesion and are hence termed easy-to-clean surfaces. Superhydrophobic surfaces have been very popular since the publication of the so-called Lotus Effect [3] in the 1990s and are often termed self-cleaning surfaces. In this context self-cleaning means that rainwater, for example, will remove any dirt that is present without other measures being necessary.

Easy-to-clean coatings with hydrophobic properties can be produced in various ways from a technical point of view. Examples include:

- Non-stick coatings based on fluoropolymers, e.g. polytetrafluoroethylene (PTFE), as often used on pans and other kitchen utensils. Coatings based on fluoropolymers are also often oleophobic (oil-repelling), namely oil has a high contact angle on the surface. This property means that oily contaminants can be easily cleaned from these surfaces.

Volkmar Stenzel, Nadine Rehfeld: Functional Coatings

ISBN: 978-3-86630-856-5

Figure 3.1: Water drops on a lotus leaf[5] (top), nasturtium leaf[6] (middle), and superhydrophobic surface (computer animation)(bottom)[7] source: S. Sepeur[2]

- Silicone coatings as, for example, used as fouling-release coatings (namely coatings from which organisms such as algae, barnacles, etc. can be easily removed) on ships.
- Sol-gel coatings with hydrophobic building blocks. A commonly used component for this is 1H,1H,2H,2H-perfluorooctyltriethoxysilane (FTS). Such sol-gel coatings lead to surfaces with a water contact angle of about 115°. Surfaces having a contact angle in this region demonstrate a water-beading effect, which goes hand in hand with low dirt uptake and easy-to-clean properties. In these coatings the sol-gel matrix provides mechanical stability and scratch resistance[4].
- Temporary coatings for the do-it-yourself area, often based on hydrophobically modified nanoparticles. These coatings are usually applied manually by being rubbed on or sprayed on and give surfaces easy-to-clean properties for a certain time (generally between a few months and a year). Example applications include car windows, shower cabins and tiles.

The aforementioned examples relate to easy-to-clean coatings (easy-to-clean surfaces). In order to produce self-cleaning surfaces, which for example can be cleaned using pure water without the addition of cleaning agents (detergents), the so-called lotus effect can be utilised.

The lotus effect is due to a microtopography superimposed by a nanotopography on the surface, combined with hydrophobic surface chemistry. This leads to a surface that possesses a Cassie-Baxter wetting regime for water (see Chapter 2.1.2). These surfaces are characterised by a contact angle for water of > 140° and a roll-off angle (the angle relative to the horizontal at which a water droplet runs off the surface) of ≤ 10°.

Figure 3.2: Principle of "self-cleaning" surfaces using the lotus effect source: S. Sepeur[2]

Such a surface means that dirt particles on the surface can be very easily removed by water, which poorly wets the surface (see Figure 3.2).

Coatings having a lotus effect can be created via a wide variety of techniques, with most being based on 3 fundamental principles[8]:

Roughening a material having low surface energy

Here, an already hydrophobic material is provided with a microtopography and nanotopography by suitable methods. The treatment introduces a Cassie-Baxter wetting regime and the surface becomes superhydrophobic. Commonly used hydrophobic starting materials are fluorinated polymers, silicones, and non-polar hydrocarbons. Examples of such methods include:

- Treatment of PTFE ("Teflon") with an oxygen plasma. This creates water contact angles of up to 168° [9].
- Roughening a PDMS (polydimethylsiloxane) surface via laser treatment. This allows a water contact angle of > 160° and a roll-off angle of < 5° to be achieved [10].
- Provision of a coating or film surface composed of hydrophobic polymers with a suitable microstructure and nanostructure (see Chapter 2.1) [11].

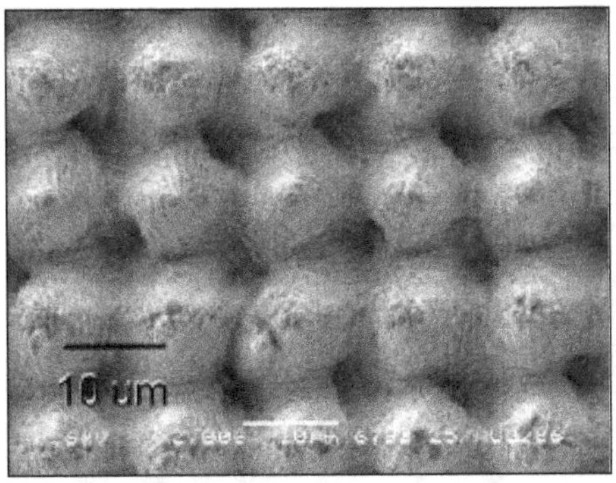

Figure 3.3: SEM image of the microstructure and nanostructure of a stainless steel surface produced by femtosecond (fs) laser pulses [14]

source: G.R.B.E. Römer, University of Twente

Manufacture of a surface with suitable roughness and subsequent modification with a material having low surface energy

Here, a suitable topography is generated in a non-hydrophobic material and a subsequent hydrophobicization step is carried out. Examples of this approach include:

- Sol-gel coatings based on various oxides (SiO_2, TiO_2, ZrO_2). By controlling the sol-gel process the coating is provided with suitable roughness. This layer is combined with suitable fluoroalkyl silanes which are used for chemical hydrophobicization [12, 13].
- Structuring of metals via treatment with ultra-short laser pulses [14]. Using this method, topographies can, for example, be produced on metal surfaces which represent a lotus-like combination of microtopography and nanotopography (Figure 3.3).

If a thin, hydrophobic, plasma-polymer coating is applied to these surfaces, one obtains a very mechanically stable, superhydrophobic surface.

Coating of a surface with a hydrophobic coating which provides a suitable microtopgraphy and nanotopography

Here, the coating already contains all the components for providing a superhydrophobic surface after being applied. Examples of such coatings include:

- Supercritical coatings based on silicone (namely having a pigment volume concentration (PVC) greater than the critical PVC). The microstructure and nanostructure here are generated by the pigments and fillers, whilst the silicone-based binder provides the hydrophobic matrix (see Chapter 2.1).
- Via plasma polymerisation. The deposition of thin layers in a vacuum process, with the layers having a suitable topography and suitable chemical composition for achieving ultrahydrophobic surfaces [15].
- Coatings based on silica microparticles and hydrophobic-functionalised polyhedral oligomeric silsesquioxane (POSS) compounds [16].
- Coatings based on microparticles and nanoparticles of silica coated with polyurethanes, which generate a combined microstructure and nanostructure. This allows surfaces to be produced having a contact angle of > 160° and an extremely small roll-off angle (0.5°) [17].

Although the lotus effect has been known for ca. 20 years, relatively few industrial applications have resulted from this, apart from some external facade paints developed about 10 years ago [18]. The reason for this is the fact that the microstructure and nanostructure can be

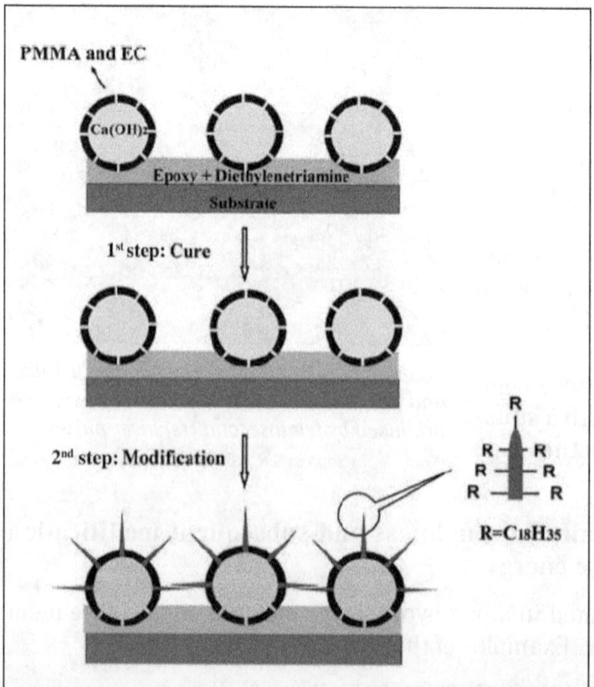

Figure 3.4: Fabrication process for regenerable superhydrophobic coatings from microcapsules (see text for details) [22]
source: Royal Society of Chemistry

very easily mechanically destroyed. This can also be observed on lotus leaves: If the leaf is touched or lightly rubbed, the beading effect, that is clearly visible on leaves that have not been touched, is lost and water shows average wetting. In nature that is unproblematic, as the leaves and surfaces can be regenerated. However, for technical applications that is generally not possible.

A further problem for the durability of the self-cleaning effect is that soap-containing or detergent-containing water no longer wets these surfaces in accordance with the Cassie-Baxter regime, meaning the desired effect is at least temporarily destroyed. If this water dries on the surface, the resulting salt and soap deposits can permanently impair the lotus effect at this site. In addition to the already mentioned factors which adversely affect the durability of the lotus effect, weathering also has to be taken into account for outside applications. In particular the effects of UV radiation, moisture, and oxygen (in some cases ozone) can lead to impairment of the effect.

Overcoming the problem of the durability of the lotus effect is a considerable challenge for researchers in order for them to develop industrial applications. Various original approaches for this have been put forward in recent years, but these are nowhere near ready for industrial implementation [19]. Examples include:

- Coatings based on hydrophobic foams [20, 21]. Here a hydrophobic, often silicone-based polymer matrix is given a foam-like structure via phase separation. Mechanical stress on such a structure does not lead to loss of the necessary microtopography and nanotopography, rather a certain number of new structures are exposed as a result of the mechanical stress.
- Regenerable superhydrophobic coatings based on permeable microcapsules [22]. With these coatings, the microtopography comprises microcapsules containing calcium hydroxide whose hulls have certain permeability for calcium ions. The capsule hulls consist of PMMA and ethyl cellulose. The capsules are secured in an epoxy matrix (step 1 in Figure 3.4). Treatment with a calcium stearate solution allows calcium stearate crystals to form on the microcapsules. These crystals are hydrophobic and form a nanotopography superimposing the microstructure (step 2 in Figure 3.4).

Such a coating has a water contact angle of up to 160°, although the roll-off angle is not less than 25° and as such this layer can only be termed superhydrophobic to a limited extent. The key feature is that this layer can be regenerated after being abraded by treating it again with calcium stearate, provided sufficient calcium hydroxide is present in the capsules.

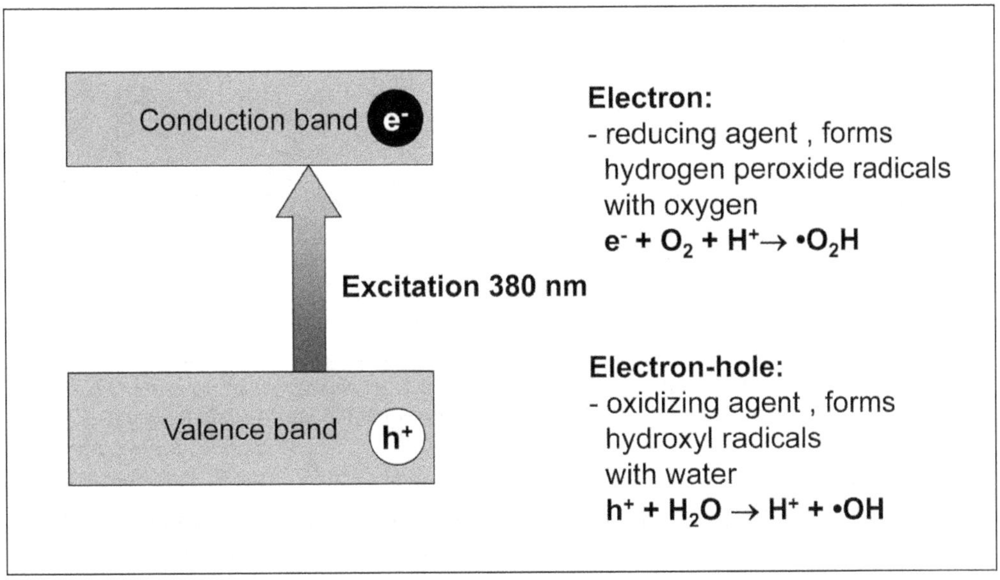

Figure 3.5: Excitation of anatase with UV light (380 nm) and chemical reactions at the surface

These approaches certainly do not represent techniques which can yet be used in industrial practice, but this work shows that there are interesting opportunities for producing stable or regenerable superhydrophobic coatings.

3.1.2 Photocatalytic surfaces

Self-cleaning surfaces based on the photocatalytic effect are based on a completely different phenomenon. Organic and oxidizable contaminants are degraded by the action of light on a suitable catalyst. In addition to the degradation of organic contaminants, photocatalytic surfaces have superhydrophilic properties, namely have an extremely low water contact angle (< 10°). Such a surface has anti-fog properties because no discrete water droplets form on the surface, but rather a thin transparent water film. A further property of photocatalytic surfaces concerns the degradation of microbial contaminants, for example bacteria and algae.

With this combination of commercially interesting properties, it is not surprising that much research work is being carried out in the area of photocatalytic surfaces.

3.1.2.1 The photocatalytic effect

The basis for the effect is photocatalysis. If a semiconductor material having a suitable band gap (gap between the conduction band and the valence band) is irradiated with light whose photon energy is greater than the band gap, then there is an electron transition within the material with formation of an electron hole pair. The electrons, and the electron holes, can diffuse to the surface and generate very reactive radicals there with suitable reactants. These radicals can in turn degrade contaminants. In addition to the surface reaction, the recombination of electrons and electron holes also takes place, in particular inside the crystals, with generation of heat.

A very effective photocatalytic material is titanium dioxide in the anatase form. Titanium dioxide can exist in an amorphous form or in various crystalline forms (anatase, rutile and brookite). Only rutile (used as a white pigment with lower photocatalytic activity) and anatase (which with its band gap of 3.2 eV is suitable as a photocatalyst) are of technical

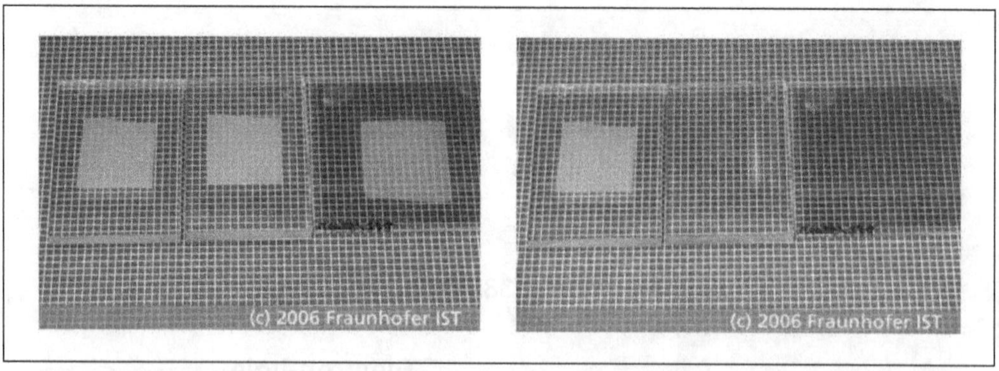

Figure 3.6: Thin layers of stearic acid on photocatalytically active samples before exposure to UV light (left image) and after 36 h exposure to UV light (0.66 mW/cm² and 366 nm)[26] (right image)

source: T. Neubert, Fraunhofer IST

importance. The band gap of 3.2 eV means that UV light of ca. 380 nm wavelength induces the described electron transition and hence the photocatalytic activity (Figure 3.5).

The radicals formed in this process with water and oxygen lead to oxidative degradation of organic and oxidizable compounds that are in contact with the surface.

As anatase has a favourable band gap, is favourably priced, and is non-toxic, it is by far the most commonly used photocatalyst for self-cleaning coatings.

Self-cleaning effect

Photocatalytic surfaces can oxidatively degrade organic contaminants using ultraviolet light. This effect has long been used to give surfaces self-cleaning properties. Japan has been the pioneer for the widespread use of this technique, but there are also some commercial products in the marketplace in Europe which utilise the photocatalytic effect, for example self-cleaning glass [23], self-cleaning concrete products [24] and self-cleaning coatings, for example for windows [25].

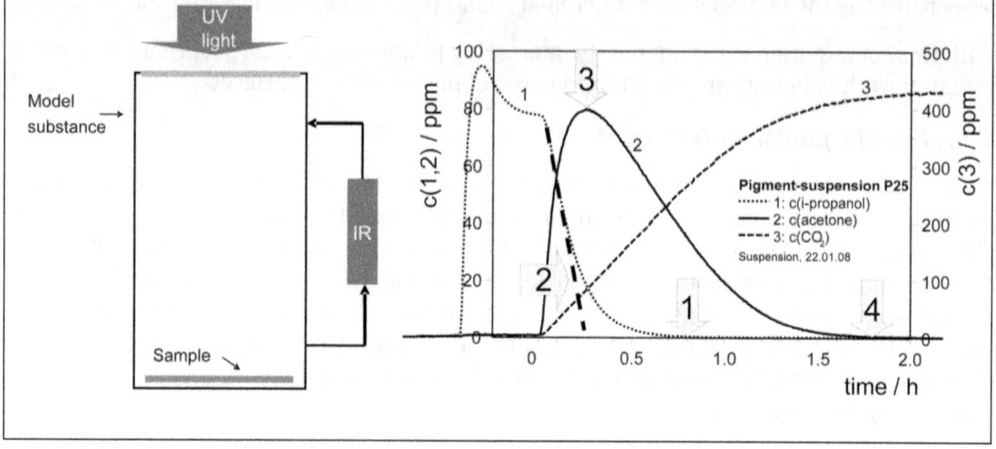

Figure 3.7: left: Reactor for measuring photocatalytic activity. Right: Photocatalytic degradation of 2-propanol on a TiO_2 layer. Arrows: 1 - time for complete degradation of 2-propanol, 2 - initial rate of decomposition of 2-propanol, 3 - maximum concentration of acetone, 4 – time for completion of all reactions[27]

source: R. Nothhelfer-Richter, Fraunhofer IPA

Various methods are used for evaluating the self-cleaning effect of surfaces. Established methods include:

- Degradation of methylene blue (standard test in accordance with DIN 52980)

In this test, two identical samples having the same active surface are placed in a methylene blue solution. One of the samples is stored in the dark, the other is exposed to a defined dose of UV-A light (e.g. 7 W/m^2). The concentration of the relevant methylene blue solutions is measured at specific intervals. Typically both samples show a decrease in the methylene blue concentration (in the sample stored in the dark this is caused by the adsorption of the dye on the sample surface). The difference between the two samples is a measure of the photocatalytic activity of the coating under study.

- Degradation of stearic acid

Stearic acid is deposited on the sample as a layer of defined thickness using a suitable vacuum chamber. The samples are then exposed to UV radiation and the degradation of stearic acid is evaluated visually or by measuring the haze (see Figure 3.6).

Other methods for analysing self-cleaning properties, and material conversion at the photocatalytic surface, include:

- Degradation of cigarette tar on white surfaces. Here, cigarette smoke is passed in a controlled way over a test surface and then the degradation of the yellow haze is carried out by exposure to UV light.
- Degradation of gaseous substances in a reactor. Here, a coated test sheet is placed in a reactor whereby it can be exposed to UV light through a quartz glass disk. The degradation of the test substance introduced into this reactor (e.g. 2-propanol) can then be followed spectroscopically. In the case of 2-propanol, one can detect the oxidation via acetone and on to CO_2 and can kinetically monitor the reaction (see Figure 3.7).

In addition to pure material conversion at the surface, the self-cleaning effect is also facilitated by the fact that the surface becomes superhydrophilic due to the exposure to light and hence has a water contact angle of < 10°. This means that water fully wets the surface, can creep below dirt and hence remove the dirt. This process is facilitated by the oxidation and hence the diminishment of the contact surface between the dirt and the surface.

Antimicrobial effect

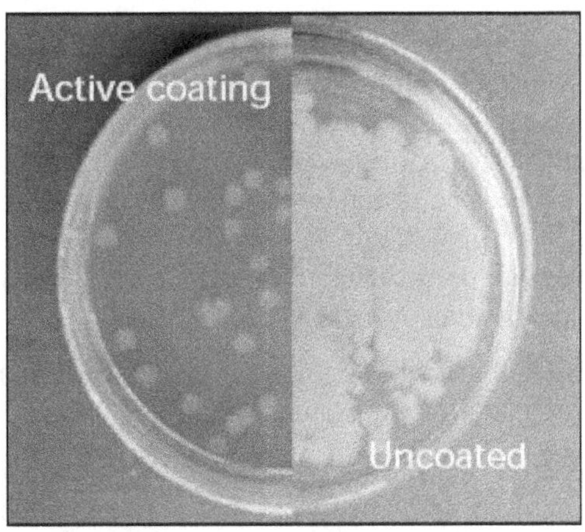

Figure 3.8: Reduction of cell growth of the bacteria E. coli caused by contact with a photocatalytically active surface[29]

source: I. Trick, Fraunhofer IGB

In addition to the self-cleaning effect, the antimicrobial action of photocatalytic surfaces is of major importance. Bacteria, algae and fungi can be deactivated and even destroyed (see Figure 3.8). Once again here it is reactive molecules generated by exposure to UV light which attack the organisms. Due to this effect it is possible to generate self-sterilizing surfaces, for example for medical applications. In Japan, photocatalytic tiles have been developed which can be almost completely self-sterilized after one hour of exposure to UV light (4 $\mu W/cm^2$)[28]. Compared to other methods, photocatalytic surfaces have the advantage that they function in a passive way, namely only UV light, moisture and oxygen are necessary. The wide use of

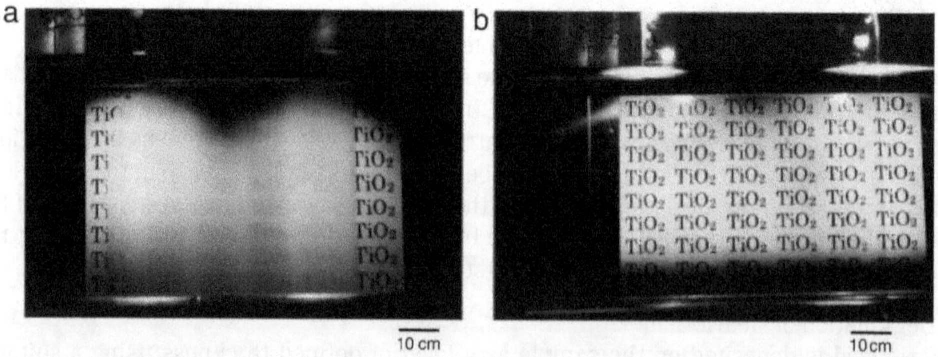

Figure 3.9: Photographs of the fogged surfaces of (a) normal glass and (b) photocatalytically coated glass after UV exposure [30]

source: Elsevier B.V.

such surfaces in hospitals, old-age homes, etc. can certainly considerably reduce the use of energy and disinfecting agents.

Anti-fog effect

The afore-described superhydrophilicity which arises on irradiating photocatalytic surfaces is used for manufacturing anti-fog surfaces. Figure 3.9 shows the effect of such surfaces on window glass compared to a normal glass surface.

Limitations

In order for the photocatalytic effect to be utilised for the manufacture of self-cleaning surfaces, four conditions must be met:

- The presence of light of a suitable wavelength
- The presence of oxygen and (atmospheric) moisture
- The presence of an active photocatalyst surface
- The contamination must be organic in nature and must be able to be oxidised

From a theoretical point of view it can be assumed that the photocatalytic effect is maintained for as long as required until any contamination is decomposed. In practice this is unfortunately not quite so because the catalyst can be deactivated by environmental factors. Volatile silicon-containing compounds (VSCCs) are a major issue [31] and are omnipresent in inside rooms. This is due to the widespread use of silicone-containing sealants, cleaning agents, deodorants, shampoos, additives in printing inks, etc. These compounds and their inorganic degradation products can cover the active surface of a photocatalyst and this can lead to slow deactivation over the course of time.

Inorganic contaminants cannot generally be removed photocatalytically and they impair the self-cleaning properties of surfaces if they are not, for example, washed off by rainwater.

A study on the long-term effect of photocatalytic glass surfaces under real weathering conditions demonstrated that although such glass surfaces become as contaminated with inorganic particles as conventional glass surfaces, the contamination of the photocatalytic glass surfaces proceeds much more slowly [32]. Such surfaces cannot therefore keep glass clean forever, but they can significantly prolong the time intervals between cleaning and so reduce the cost, energy input, and usage of chemicals for the cleaning.

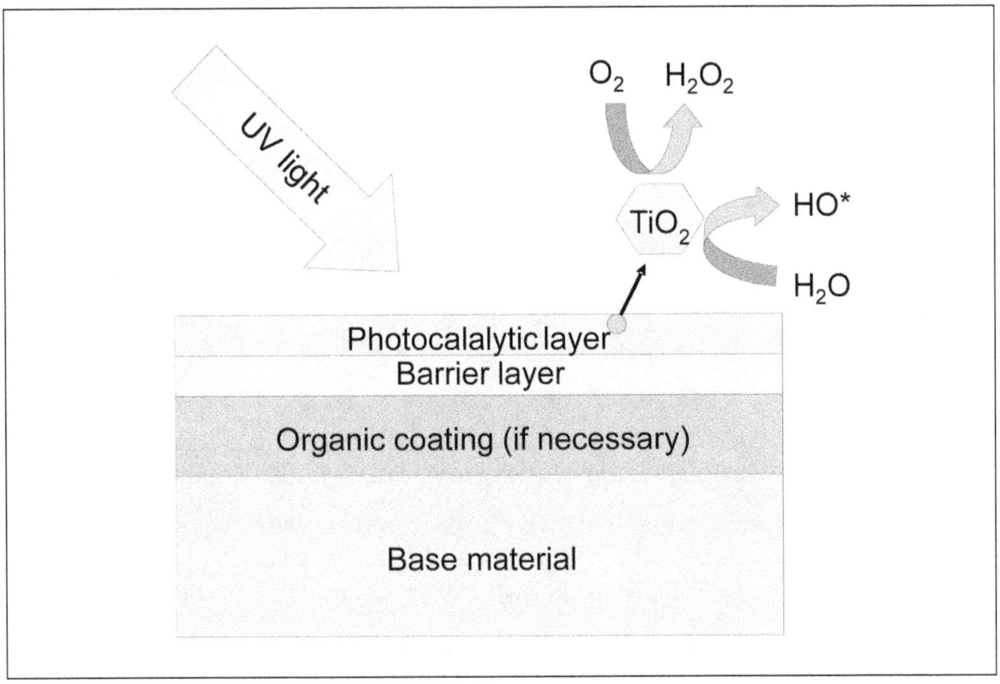

Figure 3.10: Layer structure for photocatalysis on organic substrates

The application in inside rooms is limited by the fact that the quantity of UV light that penetrates through the glass into the room is very low, and hence so is the photocatalytic effect. Development work is ongoing to modify TiO_2-based photocatalysts by doping them with suitable foreign atoms (e.g. carbon or nitrogen) such that a photocatalytic effect occurs in the visible light region, namely without UV light [33, 34].

Photocatalysts and photocatalytic layers can be readily applied to inorganic substrates such as concrete, ceramics, and glass because the reactive radicals generated in the layer cannot attack inorganic substrates. On organic surfaces (e.g. plastic, paint, wood), the radicals naturally cannot distinguish between the organic dirt and the organic substrates and indeed attack both equally. Therefore, in order to utilise the photocatalytic effect on organic surfaces, suitable (mostly inorganic) intermediate layers must first be applied (see Figure 3.10).

These intermediate layers must fulfil the following functions:

- Protection of the substrate from oxidative attack by the outer photocatalytic layer;
- Adhesion promotion between the outer photocatalytic layer and the substrate;
- Protection of the photocatalytic layer, where necessary, from components migrating from the substrate (e.g. additives migrating from plastics).

For special applications there are already systems available for the manufacture and application of such intermediate layers (see in the following). Systems for broader industrial application are the focus of current R&D work.

3.1.2.2 *Manufacture of photocatalytic coatings*

Photocatalytic coatings can be manufactured in a variety of ways. These include classical thin layer methods such as reactive pulse magnetron sputtering [35], pulsed laser deposition (PLD) [36], and atomic layer deposition (ALD) [37].

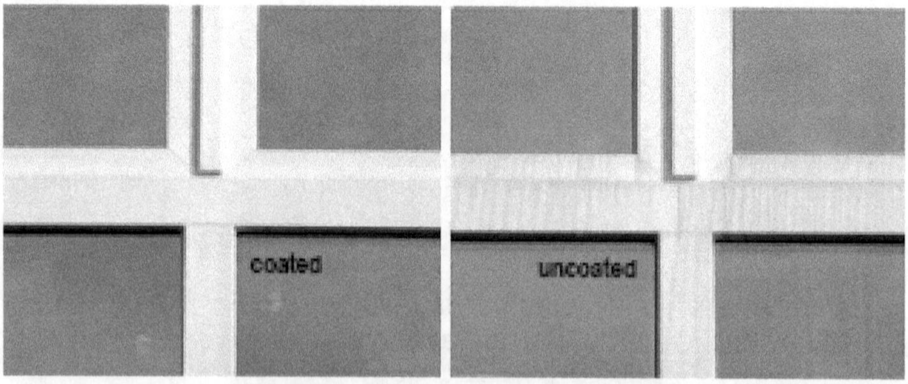

Figure 3.11: PVC window frames after being weathered for 2 years. Left: With photocatalytic coating; Right: Without coating[49]
source: F. Groß, S. Sepeur[2]

For paint or paint-like applications, the following methods come into consideration:

Photocatalytic coatings based on dispersed particles in binders

Photocalytically active pigments are commercially available in various types and in larger quantities [38, 39]. The particle size has a decisive effect on the photocatalytic activity. The optimum particle size for anatase-based photocatalysts takes into account the light absorption, light scattering, charge-carrier dynamics, and surface area and lies between 25 nm and 40 nm [40].

In order to manufacture stable photocatalytic coatings the following conditions must be fulfilled:

- The binder must be sufficiently resistant to oxidative attack by the reactive radicals. For this reason, binders with, for example, a high inorganic content such as silicones, fluorosilicone emulsions [41] and silicate-containing binders [42] are suitable.
- The photocatalytic nanoparticles, like other nanoparticles, must be dispersed and stabilised using suitable methods. Due to the extremely large surface area of the nanoparticles, they have a high tendency to reagglomerate. Studies on the dispersion and stabilisation of TiO_2 nanoparticles have demonstrated that suitable aggregates for dispersion and also suitable dispersion additives are available in order to manufacture stable dispersions [43]. In certain cases, ultrasound dispersion in combination with special gemini-surfactants or anionic surfactants have proven to be particularly suitable for producing stable, nigh agglomerate-free dispersions [44].
- The photocatalyst particles on the paint surface must be readily accessible (must not be covered with binder). This is the reason why such paints often only show their full effectiveness after a certain "run-in" period.

Paints that function as photocatalytic self-cleaning coatings (some of which also have air-purifying properties due to the decomposition of hazardous substances) are already commercially available from a variety of manufacturers. These can be used for internal and external applications for buildings, and some of these systems contain modified photocatalysts for use under visible light [45-47].

Although photocatalytic coatings have been used for buildings for many years, industrial applications are only in their infancy.

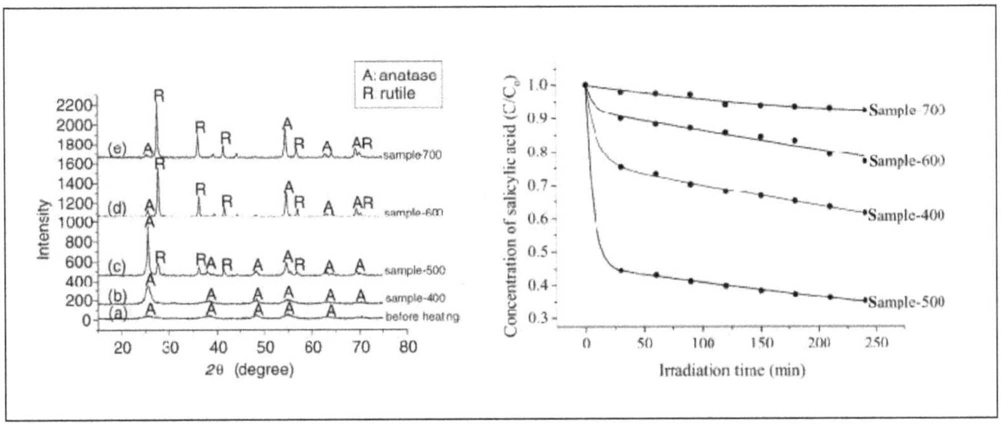

Figure 3.12: Left graph: X-ray diffraction (XRD patterns of TiO_2 materials produced by the sol-gel method (the numbers indicate the calcination temperature, e.g. sample-700 means calcined at 700 °C). Right graph: Photocatalytic degradation of salicylic acid (concentration versus irradiation time with UV light) [50]

Source: Elsevier B.V.

Photocatalytic surfaces based on sol-gel materials

The sol-gel method (see Chapter 2.4) is most suitable for manufacturing photocalatytic layers because titanium dioxide layers themselves and also intermediate layers can be prepared with a high inorganic content using this approach.

Figure 3.11 shows an example of a photocatalytic layer system on a PVC window frame. Here a silicate intermediate layer, prepared using the sol-gel method, is used as base material for a layer of available photocatalytic particles on the organic PVC substrate [48].

The images highlight how the photocatalytic layer is able to reduce the tendency of the substrate to become dirty and contaminated. Such a layer can naturally also be applied to thermoplastics and other temperature-sensitive substrate materials.

This is however not the case for the manufacture of photocatalytic titanium dioxide layers via a sol-gel reaction because these require a thermal treatment (calcination) at temperatures between 350 °C and 600 °C in order to generate a stable layer with a high content of photocatalytically active anatase. For temperature-insensitive substrates, however, the sol-gel method can be used to make very effective coatings. The starting materials are typically titanium alkoxy compounds such as titanium (IV) n-butoxide ($Ti(O\text{-}Bu)_4$) or titanium (IV) i-propoxide ($Ti(O\text{-}iPr)_4$). The reaction takes place via hydrolysis and condensation steps in accordance with the following overall equation:

$$Ti(OR)_4 + 2H_2O \rightarrow TiO_2 + 4ROH$$

Decisive for the photocatalytic effect are the reaction conditions such as the temperature and duration of the calcination step which determine parameters such as the crystal structure (whether amorphous, anatase or rutile) and the particle size and shape.

The following example shows the influence of calcination temperature on the crystal structure (Figure 3.12, left) and on the photocatalytic effect [50]. Salicylic acid was chosen as model substance for the photocatalytic degradation reaction (Figure 3.12, right).

The left-hand graph clearly illustrates that freshly prepared TiO_2 (a) is largely amorphous with only a small amount of anatase. Up to a calcination temperature of 500 °C (c), the anatase content increases, whilst above 500 °C rutile starts to form and this clearly dominates at 700 °C (e).

Figure 3.13: SEM micrograph of a TiO_2 layer on a polyurethane coating at different magnifications
source: Fraunhofer IFAM

As expected, the experiments on the photocatalytic degradation (salicylic acid as model contamination) show optimum effectiveness for a calcination temperature of 500 °C, namely when the anatase content is at a maximum.

Photocatalytic layers manufactured using the sol-gel method are hence only suitable for substrates that are not very temperature-sensitive such as ceramics and glass.

Lastly, it has also been demonstrated that sol-gel photocatalytic layers can also have an anti-reflection function (in addition to their self-cleaning function) if suitable porosity is generated [51]. These are therefore truly multifunctional layers.

Photocatalytic surfaces based on liquid phase deposition (LPD) of TiO_2

A different method for manufacturing photocatalytic layers may have advantages when working with temperature-sensitive substrates. Various methods for depositing thin, ceramic layers from aqueous solution at moderate temperatures (25 to 100 °C) with layer thicknesses between 100 and 1000 nm have been described [52].

The liquid-phase deposition of titanium dioxide layers is based on hydrolysis of a water-soluble fluoro complex (e.g. ammonium hexafluorotitanate) in the presence of boric acid:

$$[TiF_6]^{2-} + n\,H_2O \rightleftharpoons [TiF_{6-n}(OH)n]^{2-} + n\,HF$$

$$[Ti(OH)_6]^{2-} + 2\,H^+ \rightarrow TiO_2 + 4\,H_2O$$

The boric acid serves to shift the equilibrium of the hydrolysis reaction to the right side due to the removal of fluoride ions:

$$H_3BO_3 + 4\,HF \rightarrow HBF_4 + 3\,H_2O$$

The layers produced in this way also have a high content of amorphous titanium dioxide, which via calcination at temperatures of up to 500 °C can be considerably reduced [53]. However, if the reaction is carried out under suitable conditions, such layers already have a significant anatase content at lower temperatures (50 °C) without a subsequent calcination step being necessary [54].

The application of titanium dioxide layers to polymer substrates using the LPD method has been studied in detail. Layers have, for example, been successfully deposited on polyethylene terephthalate (PET), polystyrene (PS), polycarbonate (PC), and polyethylene (HDPE), with the adhesion of the layers being very dependent on the pretreatment [55].

With suitable pretreatment, cross-cut values of 0 can be attained for these substrates.

Own studies on the deposition of titanium dioxide layers on paint surfaces (2-component polyurethane) have shown that layers with good adhesion can be obtained under suitable reaction conditions. For example, layers with a cross-cut value of 0 can be attained without pretreatment. Figure 3.13 shows the topography of such a surface.

The use of the LPD method to manufacture photocatalytic layers is still in its infancy. However, it is a very promising method for coating temperature-sensitive polymer materials.

3.1.3 Literature

[1] Forschungsagenda Oberfläche: Analyse des Innovations- und Nachhaltigkeitspotenzials im Bereich der Oberflächenbehandlung, DFO Service, 2006

[2] *Stefan Sepeur*, Nanotechnology - Technical Basics and Applications, Hanover: Vincentz Network 2008

[3] *Barthlott, W.*: Scanning electron microscopy of the epidermal surface in plants, Systematics Association's Special, 41, 1990, 69-94

[4] DE 19917367 A1, "Verfahren zur Herstellung von Überzügen auf Basis flourhaltiger Kondensate", 19.10.2000

[5] http://de.wikipedia.org/wiki/Bild:Tropaeolum-majus%28Lotus-oben29.jpg

[6] http://de.wikipedia.org/wiki/Bild:LotusEffekt1.jpg

[7] http://de.wikipedia.org/wiki/Bild:LotusEffekt2mq.jpg

[8] *Ma, M.; Hill, R. M.*, Superhydrophobic surfaces, Current Opinion in Colloid & Interface Science 11, 2006, 193-202

[9] *Shiu, J.-Y.; Kuo, C.-W.; Chen, P.*, Fabrication of tunable superhydrophobic surfaces, Proceedings of SPIE - The International Society for Optical Engineering, vol. 5648, 2005, 325-332

[10] *Jin, M.H.; Feng, X.J.; Xi, J.M.; Zhai, J.; Cho, K.W.; Feng, L. et al.*, Superhydrophobic PDMS surface with ultra-low adhesive force, Macromol. Rapid Commun, 26, 2005, 1805-1809

[11] *Sun, M.H.; Luo, C.X.; Xu, L.P.; Ji, H.; Qi, O.Y.; Yu, D.P. et. al.*, Artificial lotus leaf by nanocasting, Langmuir, 21, 8978-8981

[12] *Shang, H.M.; Wang, Y.; Limmer, S.J.; Chou, T.P.; Takashi, K.; Cao, G.Z.*, Optically transparent superhydrophobic silica-based films, Thin Solid Films 471, 2005, 37-43

[13] *Taurino, R.; Fabbri, E.; Messori, M.; Pilati, F.; Pospiech, D.; Synytska*, A., Facile preparation of superhydrophobic coatings by sol-gel process, Journal of Colloid and Interface Science, 325, 2008, 149-156

[14] *Römer, G.R.B.E.; Huis in't Veld, A.J.; Meijer, J.; Groenendijk. M.N.W.*, On the formation of laser induced self-organizing nanostructures, CIRP Annals - Manufacturing Technology 58, 2009, 201-204

[15] Patentanmeldung DE 10047124

[16] *Dodiuk, H.; Rios, P.F.; Dotan, A.; Kenig, S.*, Hydrophobic and self-cleaning coatings, Polymers for advanced technologies, 18, 2007, 746-750

[17] *Su, C.*, Facile fabrication of a lotus-effect composite coating via wrapping silica with polyurethane, Applied Surface Science 256, 2010, 2122-2127

[18] www.lotusan.de

[19] *Roach, P., Shirtcliffe, N.J.; Newton, M.I.*, Progress in superhydrophobic surface development, Soft Matter, 4, 2008, 224-240

[20] *Yabu, H.; Takebayashi, M.; Tanaka, M.; Shimomura, M.*, Superhydrophobic and lipophobic properties of self-organized honeycomb and pincushion structures, Langmuir, 21 (8) 2005, 3235-3237

[21] *Han, J.T.; Xu, X.; Cho, K.*, Diverse access to artificial superhydrophobic surfaces using block copolymers, Langmuir, 21(15), 2005, 6662-6665

[22] *Wang, Q.; Li, J.; Zhang, C.; Qu, Z.; Liu, J.; Yang, Z.*, Regenerative superhydrophobic coating from microcapsules, Journal of Materials Chemistry, 20, 2010, 3211-3215

[23] www.pilkington.com/International+Products/Activ/

[24] www.saw-ch/index.php?TPL=10208

[25] www.nano-x.de/Uebersicht/#photokatalyse

[26] www.photokatalyse.fraunhofer.de/publica/publikationenAPK.asp - Measurement of photocatalystic activity with evaporated stearic acid

[27] *Christ, U.; Nothhelfer-Richter, R.; Riegert, I.; Wanner, M.; Öchsner, W.*, Ohne Bewitterung keine Photoaktivität - Optimierung der Reaktormethode zur Bestimmung der photokatalytischen Aktivität von Bautenfarben, Farbe und Lack, 11, 2009

[28] *Fujishima, A.; Zhang, X.; Tryk, D.A.*, TiO_2 photocataysis and related surface phenomena, Surface Science Reports 63, 2008, 515-582

[29] www.photokatalyse.fraunhofer.de/publica/publikationenAPK.asp - Biological efficiency measurements for photocatalysts

[30] *Watanabe, T.; Nakajima, A.; Wang, R.; Minabe, M.; Koizumi, S.; Fujishima, A., Hashimoto, K.*, Photocatalytic activity and photoinduced hydrophilicity of titanium dioxide coated glass, Thin Solid Films, 351, 1999, 260-263

[31] *Paz, Y.*, Application of TiO2 photocatalysis for air treatment: Patents' overview, Applied Catalysis B, Environmental, 2008, doi:10.1016/j.apcatb.2010.05.011

[32] *Chabas, A.; Gentaz, L.; Lombardo, T.; Sinegre, R.; Falcone, R.; Verità, M.; Cachier, H.*, Wet and dry atmospheric deposition an TiO2 coated glass, Environ. Pollut. 2010, doi:10.1016/j.envpol.2010.04.003

[33] *Dunnil, C.W.H., Aiken, Z.A.; Pratten, J., Wilson, M.; Morgan, D.J., Parkin, I.P.*, Enhanced photocatalytic activity under visible light in N-doped TiO_2 thin films produced by APCVD preparations using t-butylamine as nitrogen source and their potential for antibacterial films, Journal of Photochemistry and Photobiology A: Chemistry 207, 2009, 244-253

[34] *Ren, W.; Ai, Z.; Jia, F.; Zhang, L.; Fan, X.; Zou, Z.*, Low temperature preparation and visible light photocatalytic activity of mesoporous carbon-doped crystalline TiO_2, Applied Catalysis B: Envirinmental, 69 (39), 2007, 138-144

[35] www.photokatalyse.fraunhofer.de/publica/publikationenAPK.asp - Decomposition ability and bioactivity of photocatalytic TiO_2 based layers by reactive pulse magnetron sputtering

[36] *Farkas, B.; Budai, J.; Kabalci, I.; Heszler, P.; Geretovszky, Z.*, Optical characterization of PLD grown nitrogen-doped TiO_2 thin films, Applied Surface Science 254, 2008, 3484-3488

[37] *Pore, V.; Ritala, M.; Leskelä, M.; Areva, S.; Järn, M.; Järnström, J.*, H2S modified atomic layer deposition process für photocatalytic TiO_2 thin films, Journal of Materials Chemistry, 17, 2007, 1361-1371

[38] www.aerosil.com/product/aerosil/zh/effects/photocatalyst/pages/default.aspx

[39] www.sachtleben.de/inlude/3_1_2_0_EN.html

[40] *Almquist, C. B.; Biswas, P.*, Role of synthesis method and particle size of nanostructured TiO_2 in its photoactivity, Journal of Catalysis 212, 2002, 145-156

[41] *Wang, R.-M., Wang, B.-Y.; He, Y.-F.; Lv, W.-H.; Wang, J.-F.*, Preparation of composited Nano-TiO_2 and ist application on antimicrobial and self-cleaning coatings, Polym. Adv. Technol., 21, 2010, 331-336

[42] *Bennani, J.; Dillert, R.; Gesing, T.M., Bahnemann, D.*, Physical properties, stability, and photocatalytic activity of transparent TiO_2/SiO_2 films, Separation and Purification Technology 67, 2009, 173-179

[43] *Yaremco, Z.M.; Tkachenko, N.H.; Bellmann, C.; Pich*, Redispergation of TiO_2 particles in aqueous solutions, Journal of Colloid and Interface Science, 296(2), 2006, 565-571

[44] *Veronovski, N.; Andreozzi, P.; La Mesa, C.; Sfiligoj-Smole, M*, Stable TiO_2 dispersions for nanocoating preparation, Surface & Coatings Technology 204, 2010, 1445-1451

[45] www.stoclimasan-color.de/evo/web/sto/27762_DE

[46] www.caparol.ch/Portals/_ch/upload/images/informationsmaterial/TechForum_1_07CH.pdf

[47] www.picada-project.com

[48] Nano-X GmbH, Patent DE 101 58 433.4-43

[49] *Groß. F.*, Nano-X GmbH, Proceedings Novel Biocide Technology III, February 2007, Berlin

[50] *Su, C.; Hong, B.-Y.; Tseng, C.-M.*, Sol-gel preparation and photocatalysis of titanium dioxide, Catalysis Today, 96, 2004, 119-126

[51] *Prado, R.; Beobide, G.; Marcaide, A.; Goikoetxea, J.; Aranzabe, A.*, Delevopment of multifunctional sol-gel coatings: Anti-reflection coatings with enhanced self-cleaning capacity, Solar Energy Materials & Solar Cells 94, 2010, 1081-1088

[52] *Niesen, T. P.; Guire, M.R.*, Review: deposition of ceramic thin films at low temperatures from aqueous solutions, Solid State Ionics 151, 2002, 61-68

[53] *Yan-tao, H.; Shu-qing, S.; Zheng-ying, X.; Chun-yan, D.; Yan, M.*, TiO_2 thin films prepared from aqueous solution and their sterilizing capability, Journal of Ceramic Processing Research, 7 (1), 2006, 49-52

[54] *Masuda, Y.; Saito, N.; Hoffmann, R.; De Guire, M.R.; Koumoto, K.*, Nano/micro-patterning of anatase TiO_2 thin film from an aqueous solution by site-selective elimination method, Science and Technology of Advanced Materials, 4, 2003, 461-467

[55] *Mallak, M.*, Beschichtung planarer Substrate durch Flüssigphasenabscheidung von Titandioxid, Dissertation, Würzburg, 2006

3.2 Anti-icing coatings

Figure 3.14: Potential applications for anti-icing coatings: Cars, heat exchangers of cooling plants, and wind turbines source: Fraunhofer IFAM

In many technical fields it is desirable to prevent the icing of surfaces. This would allow, for example, the operation and reliable functioning of engines and structures in cold climates. The de-icing procedures which are currently employed involve high energy input in the form of heat, chemicals, and mechanical action. Therefore, preventing surface icing or at least reducing ice formation is also of great interest for economic reasons.

Effective anti-icing technologies would not only improve the functioning of, for example, aircraft, cars and trains but would also minimise adverse effects on bridge structures, transmission lines, and wind turbines. Furthermore, the prevention of ice formation on, for example, cooling units would save a great amount of energy. These few examples serve to highlight the importance of effective anti-icing measures.

Anti-icing coatings have been the focus of R&D work for many years. A variety of strategies have been pursued and these will be described in this section. The different types of ice and the physico-chemical properties of ice that relate to de-icing and anti-icing technologies will be covered. Current tests that are used will also be outlined.

3.2.1 Ice and its properties

Ice is the generic name for all microscopic and macroscopic forms and modifications of solid water. Ice consists of transparent crystalline structures whose nature depends on the ambient conditions. Water freezes at 0 °C under standard environmental conditions. Water in the environment contains dissolved minerals and colloidal particles (e.g. dust particles, grains of sand, aerosols) and these serve as crystallisation nuclei. Furthermore, rough surfaces can initiate crystallisation. If no crystallisation nuclei are present, water can remain in a liquid state down to -40 °C (supercooled water). Ice formation is dependent on the temperature and atmospheric pressure: the formation of stable ice crystals at -5 °C requires about 50,000 water molecules whereas at -20 °C only a few

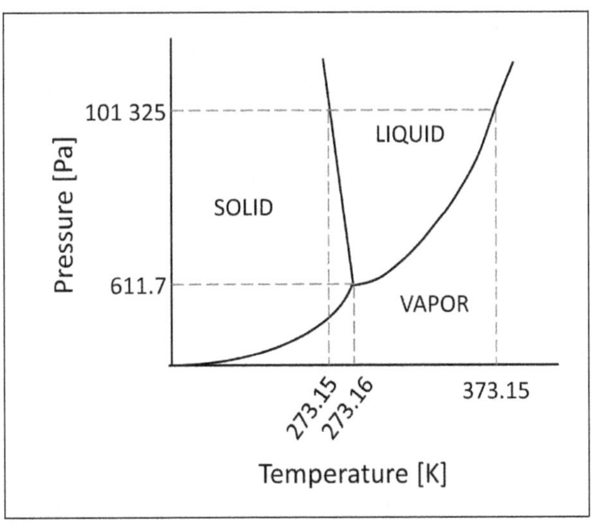

Figure 3.15: Ice formation as a function of temperature and pressure

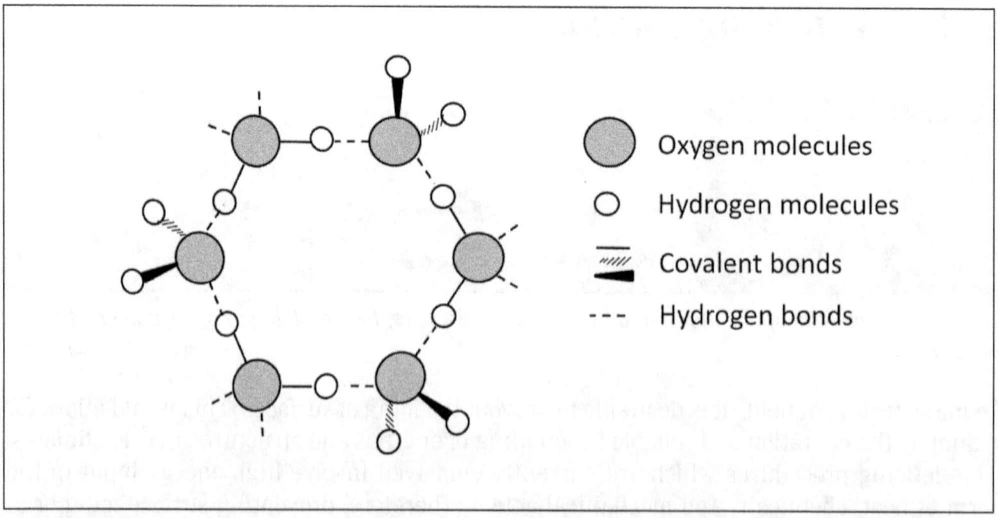

Figure 3.16: Simplified representation of water molecules and the bond types within an ice crystal

hundred molecules are necessary [1]. Ice formation also depends on the air pressure, as is clear from Figure 3.15. As standard air pressure conditions (average sea level pressure = 1013.25 hPa) often prevail for technical applications where icing is an issue, this parameter plays a secondary role.

"Conventional" ice has a hexagonal crystal structure, termed I_h. Six water molecules are connected to each other by hydrogen bonds and form a ring. This is schematically shown in Figure 3.16. Each molecule is also connected to two neighbouring rings of molecules. This structure means there is 3-dimensional long-range order and this is responsible for the typical properties of ice.

The appearance of ice varies considerably depending on the climatic conditions. Typical representatives are snow, rime, hail, and freezing rain. Whilst snow is water frozen in the atmosphere, rime forms on surfaces that are below the freezing point of water molecules present in the air (resublimation). Wind can promote rime formation. The contact of supercooled water with surfaces also causes crystallisation and hence ice formation. Clear ice is another type of ice of relevance for technical applications. This forms due the freezing of water drops (e.g. rain) on surfaces that are at temperatures below 0 °C. These different ice formation mechanisms must be taken into account when selecting a suitable anti-icing technology.

3.2.1.1 Physical properties of ice

Ice is less dense than liquid water and is therefore always present on top of the water phase. The hydrogen bonds in ice lead to a regular structure of water molecules in a crystal lattice. The distance between the water molecules is defined by the hydrogen bonds and is greater than the intermolecular distance in liquid water.

With regards to the bonds at the surface of ice crystals, not only hydrogen bonds but also further non-covalent bond types have to be considered. These are based on the dipoles in the molecules and can be divided into three types [2]:

- dipole-dipole interactions (Keesom interactions),
- dipole-induced dipole interactions (Debye interactions),
- induced dipole-induced dipole interactions (van-der-Waals interactions or London dispersion forces)

The above-described interactions are weaker than the hydrogen bonds. However, these interactions affect the adhesion of ice on surfaces. A few examples of this are given below.

Water molecules are permanent dipoles and these dipoles can interact with surfaces in different ways. If we consider the contact between water/ice and a metallic substrate, then we can expect the hydrogen bonds to be dominant. However, dipole (water)-induced dipole (metal) interactions (due to the polarisation of the electron gas of the metal atoms) will also occur. On the other hand, the dominant bond type on glass will be hydrogen bonds because of the hydroxyl groups on the glass surface.

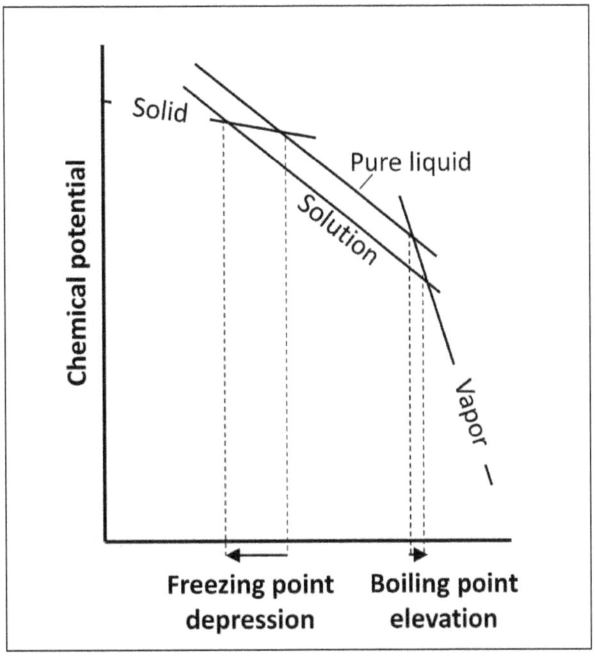

Figure 3.17: Representation of the colligative effect source:[9]

For organic coatings the bond types depend on the selected coating. In general, weak interactions (development of bonds) between the ice and coating surface and a small contact area will result in lower adhesion forces.

Hundreds of scientific studies have been carried out on ice, but only a few deal with the interactions between the ice and the solid surface (for example see ref. [3]). The interfacial properties are determined by an ultra-thin liquid layer, which covers the solid ice surface. It is called the QLL (quasi liquid layer) and was first mentioned by Faraday 150 years ago in connection with his research on ice. He pointed out that this layer is the reason for the slipperiness of ice. This explanation is disputed in other publications where this physical phenomenon is described as the effect of pressure melting or frictional heating [4].

Various studies on the QLL have shown that this layer is present down to temperatures of -35 °C or -33 °C depending on the measurement method. No QLL was detected below this, which could explain the observation that snow is "sand-like" below these temperatures. The thickness of the QLL was measured to be a maximum of 70 nm [5]. The thickness decreases with decreasing temperature and cannot be detected below -35 °C or -33 °C. Furthermore, the thickness of the QLL is dependent on other parameters:

- the higher the relative humidity of the surroundings, the thicker the layer [6],
- the presence of dissolved salts increases the thickness [7]

Experience has shown that ice adhesion decreases with decreasing temperature [8]. However, the influence of the QLL on ice adhesion is not fully understood. Nevertheless, it should be kept in mind when interpreting the adhesion mechanisms of ice on different surfaces.

3.2.1.2 Physico-chemical properties of ice

Whether a substance is in the liquid, solid or gaseous state not only depends on the temperature and pressure but also on the concentration of solutes (Figure 3.17). Ice formation can be suppressed by increasing the concentration of solutes. Here, the freezing point depression is related to the colligative properties and is determined solely by the concentration of dissolved molecules, and not their nature.

Figure 3.17 shows the correlation between the chemical potential (ordinate) and temperature (abscissa). The lines (from left to right) show the chemical potentials of the solid phase (pure substance), liquid phase (pure liquid; solution), and vapour phase. The chemical potential depends on the temperature and increases with decreasing temperature. At the freezing point, the solid phase and the liquid phase have the same chemical potential, both are energetically equivalent. The lowering of the freezing point can be described as follows: When a solute is added to a pure liquid, the chemical potential of the solution decreases, but the chemical potential of the solid phase solvent is not affected (assuming that a solute only dissolves in the liquid solvent and not in the solid solvent). This means in turn that the "intersection" of the solid and liquid phase lines occurs at a different temperature, and so the freezing point is depressed. The use of chemical substances for freezing point depression is an interesting approach and will be discussed in Chapter 3.2.2.1.

Besides the colligative freezing point lowering, it is also possible to suppress or minimise the formation of ice using the constitutive (structural) properties of anti-freeze proteins. Such proteins have already been isolated from amphibians, insects, and plants. These proteins allow fauna and flora to survive in cold climates [10]. Further information about anti-freeze proteins can be found in Chapter 3.6.3.2, where it is also explained how the properties of these special proteins can be transferred to surfaces.

3.2.2 Anti-icing/de-icing technologies

Various methods are currently available for suppressing ice formation and for removing ice from surfaces. A distinction can first of all be made between anti-icing technologies (which suppress ice formation at the initial stage) and de-icing technologies (which remove ice that has already formed). These methods can also be split into active and passive methods, as summarised in Table 3.1. The methods that are highlighted in Table 3.1 can be realised via the coating – and these are generally termed anti-icing coatings regardless of whether they function by suppressing ice formation or by facilitating the removal of ice that has already formed. Other terms used in the literature include icephobic and ice-release.

Most technologies involve active methods, with the heating of a surface (thermal) being a very effective and widely used method in many sectors of industry (e.g. automotive sector, aviation industry). Besides electrical heating methods, other heating methods can be used (e.g. bleed air, infrared radiation, microwave heating). In all cases these methods can be used for anti-icing as well as for de-icing. Thermal methods are very effective for a wide range of applications. Their disadvantage is the high and costly energy input required to heat surfaces under cold climatic conditions. For this reason, current research activities are focusing on technologies that reduce the energy costs or use alternative technologies.

Another active method is the lowering of the melting/freezing point of water by chemical means. Chemicals are employed in anti-icing solutions (e.g. for spraying aircraft with a mixture of glycols before they take off, so maintaining a protective anti-icing film) and in de-icing solutions (e.g. salt solutions on road surfaces that melt any ice that is present). The use of these

Table 3.1: Anti-icing and de-icing methods, with the coating-related methods highlighted

Active methods	Passive methods
Thermal	Reduction of ice adhesion
Chemical	
Biochemical	Change of wetting behaviour
Mechanical	

freezing point depressors in coatings is discussed in detail in Chapter 3.2.2.1, as is the use of biochemically based anti-freeze proteins for preventing ice formation. These systems and strategies are at an early stage of development.

We all know from scraping ice off our car windscreens that mechanical methods can also be used for de-icing. In addition, impulsive or expulsive electromechanical systems can also be used to remove ice from surfaces. These methods can be aided by passive methods that reduce ice adhesion and hence facilitate removal (de-icing). Another passive method is to alter the wetting behaviour of surfaces so that most of the water runs off before freezing. These passive methods can be realised using suitable coatings (see Chapter 3.2.2.2 for details).

3.2.2.1 Active anti-icing coatings

This section describes coatings that actively suppress ice formation on surfaces. These can be split into coatings that depress the freezing point of water due to their colligative properties (chemical method) and coatings that suppress ice formation due to their constitutive (structural) properties (biochemical method). Although both strategies suppress ice formation by lowering the freezing point of water, the fundamental mechanisms involved are very different. Figure 3.18 highlights the differences between the strategies.

Implementing the chemical approach in coatings initially seems to be relatively simple: Known freezing point depressors such as salts or glycols are incorporated into organic coating systems and these should depress the freezing point of water when water comes into contact with the coating surface. Studies have shown that this approach is possible, with factors such as the compatibility of the compounds and longevity of the anti-icing effect having to be considered. Figure 3.19 shows a number of model coatings containing different freezing point depressors.

The anti-icing effects that can be realised by such coatings have been investigated in a host of studies. Table 3.2 shows test results that demonstrate that both the coating matrix and the nature of the freezing point depressor are key factors. It must be stressed here that the resulting anti-icing effects are relatively short-lived. Once the freezing point depressor has been leached out of the coating there is no longer any effect.

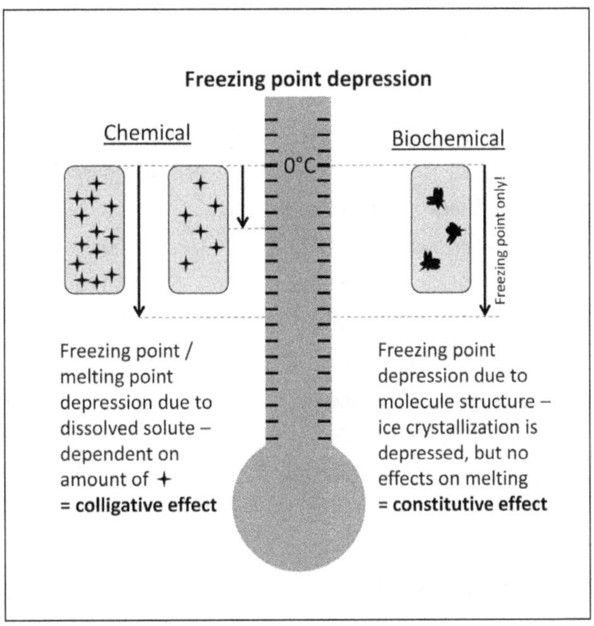

Figure 3.18: Different strategies for depressing the freezing point of water

Possible applications for these short-lived anti-icing coatings include on means of transport, in order to suppress ice formation for a specific period of time. The large number of available freezing point depressor/organic coating combinations allows customized coatings to be developed for specific technical requirements. Here the optimum balance between the longevity and effectiveness must be found. If too little anti-icing agent gets to the surface then the

Figure 3.19: Model coatings demonstrating the effect of freezing point depressors on the appearance of the coating: (1) PUR dispersion without additive, (2) PUR dispersion containing polypropylene glycol – no visible change in the appearance of the coating film, (3) PUR dispersion containing urea – significant change in the appearance of the coating film *source: Fraunhofer IFAM*

effect is too small, yet if the anti-icing agent is leached out too rapidly then the effect is too short-lived.

Table 3.2: Results of rime tests, demonstrating that the use of freezing point depressors significantly reduces rime formation. Key factors are the nature of the anti-icing additive and the type of coating matrix. Test conditions: air temperature +1 °C, substrate temperature -15 °C, relative humidity 88 %, wind speed 10 m/s (for further information about the tests see Chapter 3.2.3).

	Rime thickness after 20 min [µm]
Acrylate dispersion without additive	900
PUR dispersion without additive	900
Glycol 1 (10 %) in acrylate disp.	300
Glycol 1 (10 %) in PUR disp.	750
Glycol 2 (10 %) in PUR disp.	450
Salt 1 (10 %) in acrylate disp.	0
Salt 2 (10 %) in acrylate disp.	0
Salt 3 (10 %) in acrylate disp.	150
Salt 1; 2; 3 in PUR disp.	incompatible
Salt 4 (10 %) in acrylate disp.	30
Salt 4 (10 %) in PUR disp.	0
Salt 4 (20 %) in PUR disp.	0
Salt 4 (50 %) in PUR disp.	0

Anti-icing coatings that function by lowering the freezing point do not necessarily have to be based on leaching effects. More recent development work has been investigating the use of firmly bonded freezing point depressors to achieve the desired anti-icing effects (being carried out by Fraunhofer IFAM in collaboration with Leibniz Institute of Polymer Research Dresden, Germany). This work is focusing on polyethylene glycol chains (PEG), which have been mentioned elsewhere in this book in other contexts (see for example Chapter 2.5 and 2.6). The challenge here is to manufacture stable coatings having an acceptable longevity for technical applications. If successful, these non-leaching systems will certainly have much promise.

The same applies for anti-icing coatings involving biochemically-based lowering of the freezing point. This approach is also currently being researched and involves the use of anti-freeze proteins. Anti-freeze proteins have been identified in organisms that live in polar and sub-polar regions. The survival of fish, amphibians, plants, and insects [11] in these regions is due to these anti-freeze proteins. Indeed, poikilothermal organisms can even survive at temperatures of -40 °C and lower [12]. The anti-freeze effect here is based on substances that cause freezing point depression due to the configuration and conformation of the molecules, namely a constitutive effect. Such molecules selectively depress the freezing point, but have no effect on the melting point of ice. This leads to a temperature difference between the melting point and freezing point. This is called "thermal hysteresis" [13]. The molecules are described as "thermal hysteresis proteins" or more commonly as "anti-freeze proteins". These proteins are able to depress the freezing point some 200 to 300 times more than substances having colligative properties [14].

In order to transfer these anti-icing properties to surfaces, the following steps have to be undertaken:

- Identification of the molecules responsible for the anti-icing effect
- Isolation or synthesis of the active molecule structure
- Manufacture of suitable surfaces for immobilizing molecules
- Covalent immobilization of the molecules without activity loss

These steps are currently being researched, with the aim being to produce stable anti-icing coatings for a variety of technical applications. The work is extremely complex and more details are given in Chapter 2.6. Test coatings containing anti-freeze proteins have already demonstrated the fundamental viability of this approach (see Chapter 3.6.3.2) which describes biomimetic coatings.

With regard to the application of active anti-icing coatings, it is expected that short-lived anti-icing coatings (namely those based on a leaching mechanism) could be commercialised within a relatively short space of time. In contrast, much more development work is still required on coatings containing active components that are firmly bonded to the coating surface and so give a stable and long-lived anti-icing effect.

3.2.2.2 Passive anti-icing coatings

Over recent years much development work has been carried out on passive anti-icing coatings and de-icing coatings. Most of these coatings function by minimising the contact with water, so considerably reducing ice formation. A further goal is to minimize ice adhesion, so that the ice can be easily removed by scraping, vibration and even by the own weight of the ice. Various approaches have been and are being pursued to minimise ice adhesion and the most salient are described below.

Hydrophobic coatings

Most of the commercially available, passive anti-icing coatings form hydrophobic (namely water-repelling) surfaces. The hydrophobic nature of a surface can be quantified by measuring the contact angle of water, which depends on the interactions between water and the surface. The lower the interaction, the greater the contact angle (see also Chapter 2.1). Hydrophobic surfaces have contact angles greater than 90°, and superhydrophobic surfaces have contact angles greater than 140°.

To create these passive anti-icing surfaces not only is the hydrophobic character of the surface important but also the surface roughness and the types of bonding sites that are available for water molecules (see Figure 3.20). A few examples will clarify this:

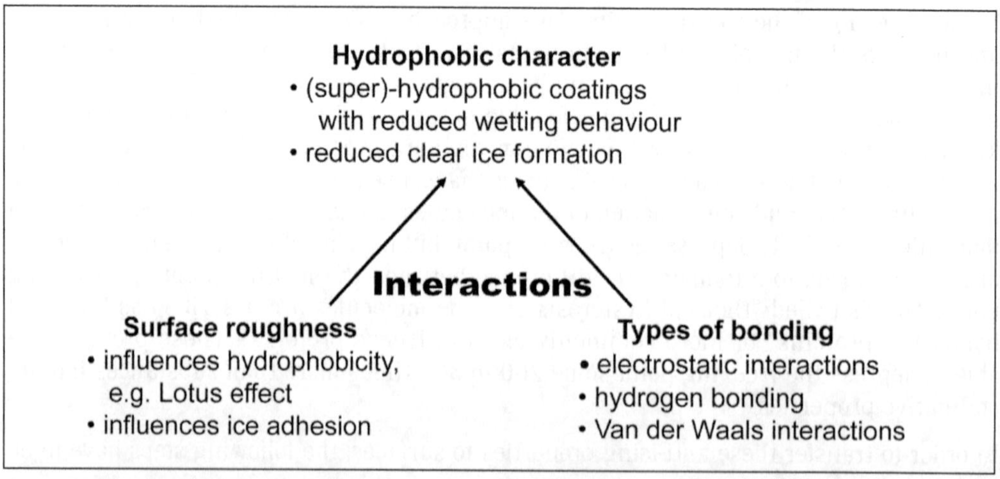

Figure 3.20: Key factors for creating passive anti-icing coatings

- In general, ice adheres more strongly to metal and glass surfaces than it does to the surfaces of many polymers.
- If there is suitable roughness, the surface hydrophobicity increases. This means that interactions with water molecules are significantly reduced (e.g. lotus effect). See also Chapter 2.1 which describes wetting according to the Wenzel regime and Cassie-Baxter regime.

These conditions do not however suffice to create passive anti-icing coatings because hydrophobic polymer surfaces are not necessarily effective anti-icing coatings. Indeed, a balance between the parameters shown in Figure 3.20 is necessary and these must be determined for each specific application.

Most hydrophobic organic coatings are based on fluorine or silicone modified materials as has already been explained in detail in Chapter 3.1 for self-cleaning surfaces. An example

Table 3.3: Comparison of an unmodified PUR coating with a fluorine-modified PUR coating that was developed as an anti-icing coating. The anti-icing effect is due to improved run-off behaviour of the water.

	Unmodified PUR coating	**F-modified PUR coating (passive anti-icing coating)**
Water contact angle [°]	82	124
	Hydrophobicity increase →	
Roughness Ra [µm]	0.17 (±0.01)	0.64 (±0.07)
	Roughness increase →	
Ice formation under freezing rain conditions: • Temp. -5 °C • 10 s rain application • Wind speed 10 m/s		
(for details see Chapter 3.2.3)	Clear ice formation drastically reduced →	

of such a hydrophobic, passive anti-icing coating is also shown in Table 3.3: The hydrophobic character of this PUR coating modified with fluorine containing additives is shown by the much higher water contact angle. The roughness value R_a (average roughness in µm) is also higher. These properties result in much reduced ice formation because (rain) water rolls off the surface faster than it does on unmodified coatings.

With regards to the above-described anti-icing coating, ice adhesion tests showed that ice adhesion is reduced by the fluorine additives (for details about the tests see Chapter 3.2.3). This is due to the lower tendency for bonding between water molecules and fluorine atoms. Countering this effect, however, is the surface roughness, because altered surface topography can lead to altered wetting behaviour and hence enhanced ice adhesion. In the worst case the increased surface roughness may permit anchoring of the water molecules. There must hence be an optimum balance between the factors indicated in Figure 3.20.

Besides organic coatings, the sol-gel coatings described in Chapter 2.4 are also used as the basis for passive anti-icing coatings. These are characterised by extraordinary hardness, mechanical stability, and excellent adhesion properties. Furthermore, modification of the inorganic components with organic substitutes leads to a combination of desirable properties involving both organic and inorganic chemistry. Our experience is that these so-called hybrid polymeric sol-gel coatings can provide very smooth surfaces with hydrophobic characteristics.

At first it might seem that hydrophobic coatings confer anti-icing functionality merely due to the reduced wetting. However, comprehensive studies have shown that the hydrophobicity is just one of many factors which determine the anti-icing properties – and furthermore a hydrophobic coating is not necessarily an anti-icing coating [15]. In addition, these passive anti-icing coatings have their limits: For example, they cannot prevent rime formation. This applies for both inorganic and organic coatings and has also been reported elsewhere [16].

Superhydrophobic coatings

Superhydrophobic surfaces have water contact angles of over 140°. The best known examples are artificial lotus surfaces. Water droplets do not wet these surfaces and forms pearls immediately (see Chapter 3.1). It is suspected that surfaces such as these may also be of interest for anti-icing coatings. As already described for hydrophobic coatings, this only applies under certain conditions: the positive effect of improved water run-off is also found here under freezing rain conditions. However, rime formation can also not be prevented here. Furthermore, no information is currently available about the effect of these surfaces on ice adhesion.

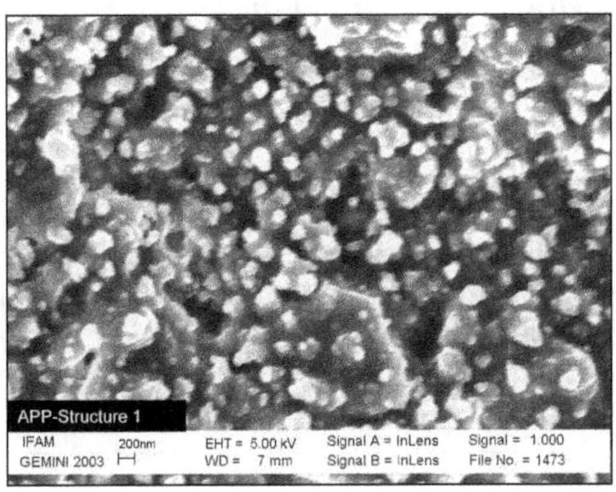

Figure 3.21: SEM image of a plasma-polymer coating with additional structures due to particle incorporation source: J. Ihde, Fraunhofer IFAM, Bremen

The manufacture of superhydrophobic surfaces has already been described in detail in Chapter 3.1. Plasma technology can also be used to generate such surfaces. Plasma itself is a partially ionized gas and can be used to deposit coatings. For example: the plasma-process can be adjusted in such a way that nanoparticle and microparticle clusters are formed, which then create a surface with a combination of nanostructures and microstructures (see Figure 3.21). These coatings have water contact angles of greater than 150°. The reduced wetting due to the artificial Lotus structure leads to reduced ice formation under freezing rain conditions. Like the coatings described in Chapter 3.1, the durability of these surfaces does not yet meet technical requirements.

Temporary (passive) coatings

Temporary coatings involve a completely different approach. Their low adhesion properties or low cohesive forces mean that ice can be more easily removed from surfaces that are already iced up. Such coatings should hence be more appositely termed de-icing coatings. In general, an additional active measure must also be taken in order to remove the ice. A variety of systems are conceivable for this approach including:

- Paste-like materials (lubricants) which prevent direct contact between the ice and substrate surface. Due to their consistency they can easily be removed together with any ice (low cohesive forces). This was reported for Li-based grease [17]. Our own studies have shown that the removal of grease with any deposited rime or clear ice required less energy compared to the removal of rime on untreated metallic surfaces.
- Wax-like substances (slip agents) that are used as temporary coatings, for example in the winter sport area (skis, etc.) to achieve good sliding properties. It is assumed here that these substances minimise the ice adhesion. Depending on the thickness of the applied substance, similar effects to the afore-described paste-like materials are also possible.
- Brittle top-coat materials which are porous due to their chemistry and/or due to the low temperatures and which are easy to remove together with any ice by mechanical means. Realisation of such a system using, for example, sol-gel based materials would be possible, because it is known that certain of these formulations tend to be very brittle.

The authors are unaware of any current industrial scale use of such temporary coatings. This is probably because these systems not only have advantages (easier ice removal, easy material handling in many cases) but also some shortcomings. The latter include a short service life and an impact on the environment, especially when coating large areas.

Hydrophobic/hydrophilic structuring

A further, relatively new approach is the development of coatings that contain hydrophilic centres in a hydrophobic environment. These coatings allow water molecules to adhere to the hydrophilic sites, while the hydrophobic area around these sites promotes the removal of ice crystals (schematically shown in Figure 3.22).

The objective here is not to suppress the ice crystallisation but rather to facilitate ice removal via a flowing medium (liquid or even an air flow). The introduction of various nanotechnologies, described elsewhere in this book, allows such domains to be realised on surfaces. The use of nanoparticles for this application is under development. A patent describes ice nucle-

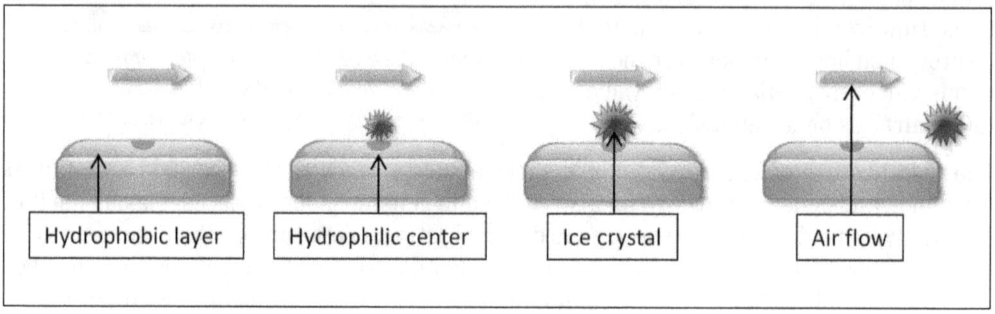

Figure 3.22: Approach for minimising ice adhesion: coatings with hydrophilic sites for ice nucleation surrounded by a hydrophobic surface. The ice is easily removed by an air flow.

ating non-stick coatings [18]. For example purposes the application of such a coating in an ice cream machine was mentioned. Here ice formation is desired, but its strong adhesion to the surface necessitates the use of expensive mechanical scraping devices or heat input in order to remove the ice cream from the machine. A heterogeneous coating that on the one hand promotes ice formation (hydrophilic domains) and on the other hand minimises ice adhesion (hydrophobic domains) may have considerable benefits compared to a homogenous surface. In order to manufacture such heterogeneous surfaces, nanoparticles are first of all synthesised using sol-gel technology and these are then incorporated into an inorganic-organic network to form these ice repellent surfaces [18].

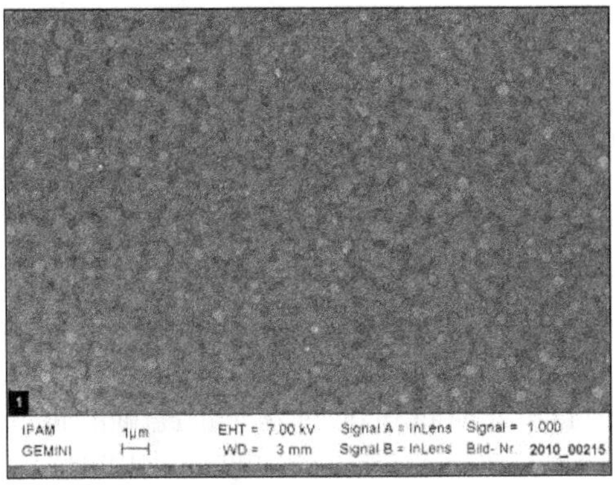

Figure 3.23: SEM image of a heterogeneous coating film, formulated using immiscible components in order to create hydrophilic/ hydrophobic domains source: Fraunhofer IFAM

Another way of creating such heterogeneous surfaces is, for example, to use coating formulations containing incompatible components. These separate during the curing process and so produce a heterogeneous surface with defined properties. The formulation must be chosen such that the desired separation of the resin components occurs, so altering the curing of the coating. Figure 3.23 shows a SEM image of such a coating system, with domains of acrylate and siloxane polymers.

Ice and rime formation was studied in detail on the resulting heterogeneous surface (for details see Chapter 3.2.3). It was found that rime only forms on part of the surface – seemingly on the hydrophilic domains (see Figure 3.24). These results confirm the effect described by Zwieg et al. (2002) [18] for promoting ice formation.

Further studies and practical trials are required in order to determine the extent to which these hydrophilic/hydrophobic domains can be used to produce effective anti-icing coatings. Specific applications, for example in ice cream machines, have already been demonstrated [18]. This lowering of ice adhesion may also be of interest for other technical applications, especially if it can be demonstrated that ice can be removed due to its own weight and in combination with an air flow.

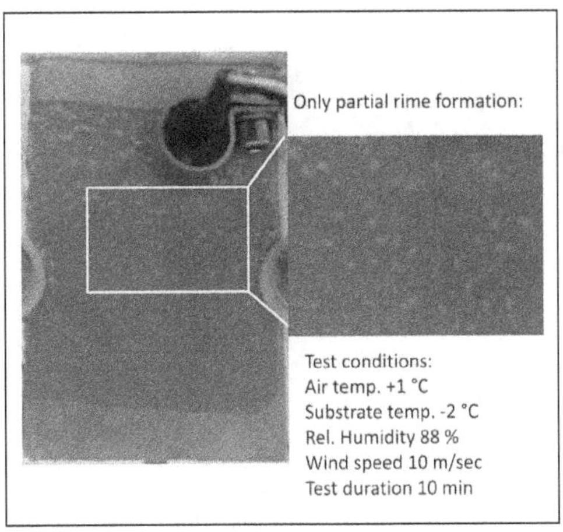

Figure 3.24: Rime formation on a coating film having hydrophilic/hydrophobic domains (for test details see Chapter 3.2.3) source: Fraunhofer IFAM

Summary

Due to their hydrophobic surfaces, passive anti-icing coatings are able to reduce ice formation and minimise ice adhesion under certain conditions. Most commercial coatings are based on this principle – but no general anti-icing effect can be achieved because the formation of rime cannot be suppressed. The effectiveness of most of these coatings has a time limit, and technical requirements are often not met. This explains why these coatings, despite much research effort, have not yet reached the marketplace.

Based on the current state of technology, it can be assumed that a universal anti-icing coating for all types of applications and effective under all icing conditions is unlikely to be available in the near future. The focus of development work will continue to be on customised coatings for specific icing conditions and specific technical requirements. We must wait and see whether new ongoing development work, such as the development of coatings with hydrophilic/hydrophobic domains, will make a contribution here.

3.2.3 Anti-ice evaluation

The icing of surfaces is an extremely complex field of research and requires special methods and equipment in order to learn about the icing mechanisms at a molecular level. It is also necessary to test the coatings under near-real conditions. This is because predictions under laboratory conditions have shortcomings due to the diverse range of icing processes. In summary this means that two types of test methods are necessary:

- Analytical methods for measuring physical and chemical parameters under well-defined laboratory conditions. These methods allow statements to be made about the mechanisms of anti-icing surfaces.
- Icing tests under near-real conditions which hence allow statements to be made about the effectiveness of anti-icing coatings and give direction for further development work.

Both types of methods require complex analytical instrumentation and there is a need for specially trained personnel to develop the anti-icing coatings. The analytical methods include microscope methods to monitor crystal formation on surfaces and differential scanning calorimetry (DSC) to measure the effect of freezing point depressors in coatings (see Figure 3.25). Other useful methods include atomic force microscopy (AFM) to study physical properties such as friction and adhesion and to detect the Quasi Liquid Layer (QLL) (see Chapter 3.2.1.1) on ice surfaces (also see ref. [31]). Surface analytical methods help us to understand the icing mechanisms on surfaces and so develop novel anti-icing coatings.

Besides the analytical measurements, it is also necessary to carry out icing tests under near-real conditions. In general refrigeration chambers or icing chambers are used for this, with the temperature, air humidity, precipitation, and wind effects being able to be controlled. Large test plants allow, for example, whole trains or other vehicles to be tested

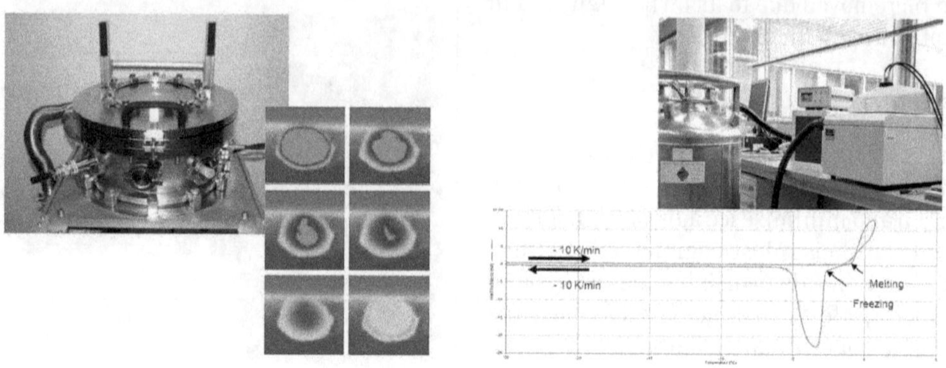

Figure 3.25: Analytical methods for studying icing on surfaces. Left: Test chamber for studying the freezing of water drops and IR images of a water drop in the process of freezing; Right: DSC instrument for studying anti-icing effects due, for example, to freezing point depressors

source: Fraunhofer IFAM

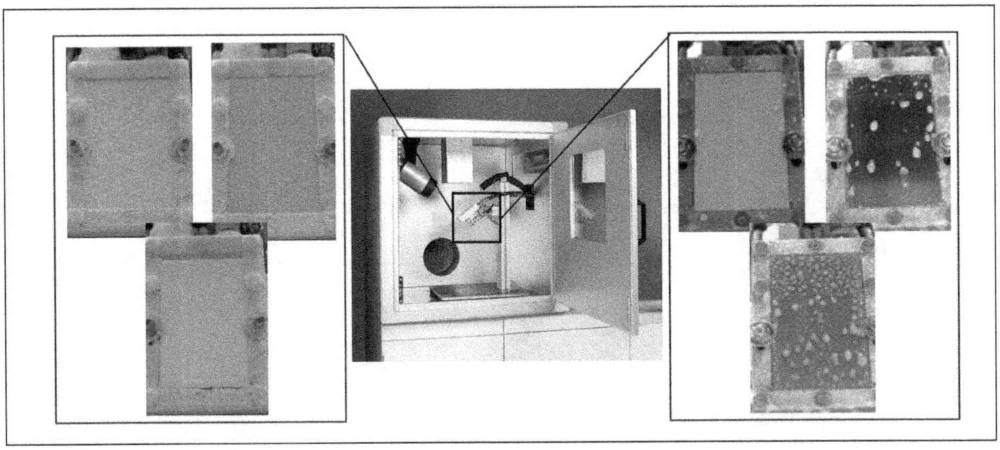

Figure 3.26: Ice chamber tests for mimicking different icing scenarios. Left: Rime formation on surfaces (conditions: air temp. +1 °C, substrate temp. -5 °C, relative humidity 88 %, wind speed 10 m/s, test duration 20 min, subsequent determination of the thickness of the rime layer); Right: Freezing rain on surfaces (conditions: temperature -5 °C, relative humidity 65 %, wind speed 10 m/s, 10 s of rain, test duration 10 min, subsequent determination of the ice formation) source: Fraunhofer IFAM

under near-real conditions (see for example www.vewip.at). Influences such as the geometry of components or the effects of de-icing methods can also be studied. The latter include, for example, the effect of runback ice formation. The ice that melts on the front edge of wind turbine blades or aircraft wings can, for example, refreeze on the back part of the blade

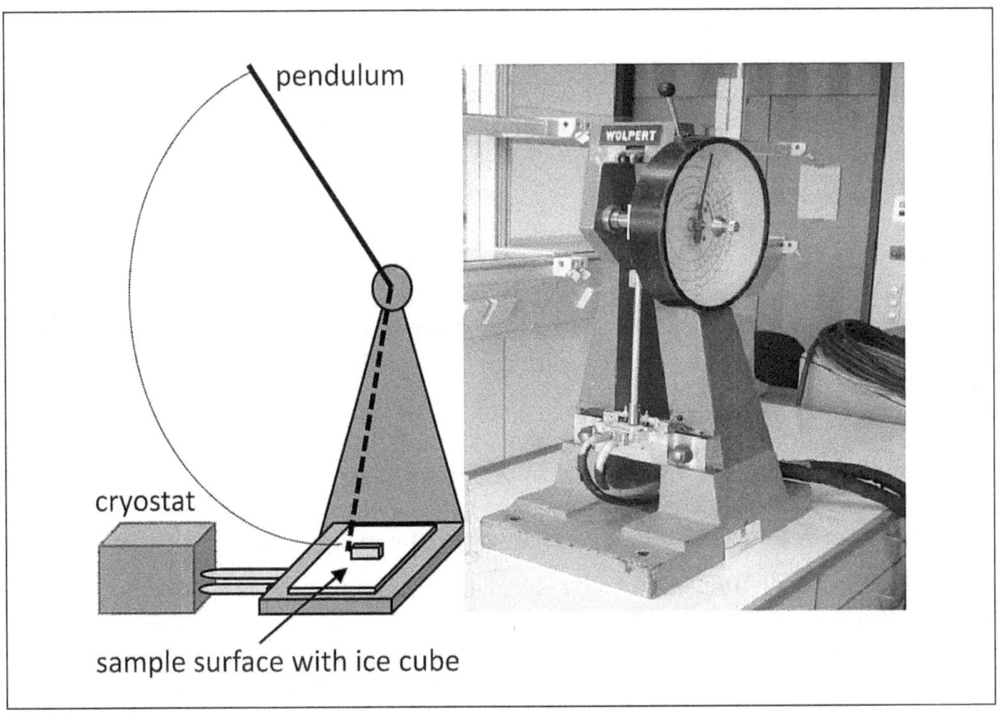

Figure 3.27: Instrument for measuring ice adhesion: Modified notch test unit, whereby the ice is knocked off the surface by a pendulum. The reduced energy of the pendulum is then correlated to the adhesive strength of the clear ice, measured as the angle of the pendulum amplitude source: Fraunhofer IFAM

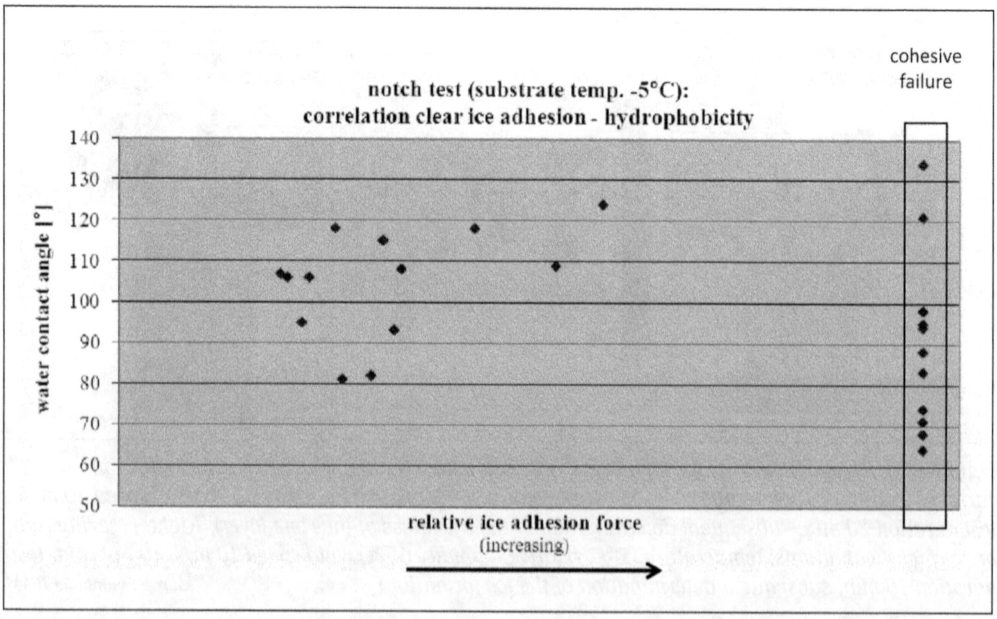

Figure 3.28: Studies on the correlation between hydrophobicity (water contact angle) and ice adhesion using the notch test on different coating types. Cohesive failure indicates situations where the adhesion of ice on the surface is stronger than the cohesion within the ice *source: Fraunhofer IFAM*

or wing and adversely affect the aerodynamic properties. This example clearly shows the importance of the tests under near-real conditions. It is however very costly to operate large test plants and it is only sensible to evaluate anti-icing coatings that are at a fairly late stage of development. Initial tests on coatings are carried out on a laboratory scale. Near-real test conditions are also a must here. Various icing scenarios must be taken into account, because good test results under, for example, freezing rain conditions do not necessarily go hand in hand with good results for rime formation. This applies for many of the passive anti-icing coatings described in Chapter 3.2.2.2. Figure 3.26 gives examples of icing tests under different icing conditions.

Besides tests on the icing of surfaces, information about the adhesion of the ice is very valuable, in particular on the passive coatings described in Chapter 3.2.2.2. A variety of test methods have been developed up until now by different work groups. These include ice adhesion tests using a centrifuge [19], torque test [20], plate shear test [16], and notch test. Figure 3.27 shows one unit that can be used for ice adhesion testing.

The ice adhesion tests are very sensitive, meaning that very small changes in the test conditions can give different results. This is why literature data often differ widely. For example, our own tests using the notch test gave no correlation between hydrophobicity and ice adhesion (see Figure 3.28). The same was found by C. Laforte et al. (2002) [17], yet other work groups have provided evidence for a correlation (see, for example, ref. [20]).

The ice tests generally provide reproducible data, meaning that it is possible to compare different coatings. As such the tests can be used for developing anti-icing coatings. However, even very small changes to the test set-up lead to different results, because the icing mechanisms are very complex. When carrying out icing tests, the following points must hence be heeded:

- the test parameters for assessing the icing behavior must be selected with care,
- even small variances can lead to misinterpretation of results,

Figure 3.29: Left: Test stand for wind turbines in Gütsch, Switzerland, source: http://www.meteotest.ch/cost727/met.html; Left: Test stand on the Brocken for testing anti-icing coatings under real conditions *source: Fraunhofer IFAM*

- the sensitivity of the icing process means that there is a relatively high risk that the test results will not correlate with the icing behavior under real conditions.

As a consequence, it is advisable to perform a broad range of tests under different conditions and then subsequently to perform tests under real conditions.

Tests under real conditions can be used to demonstrate the icing behavior of complex structures (e.g. wind turbines) taking into account all technical requirements. The tests can also be used to determine the longevity of the anti-icing effects. This is of vital importance for all end-users, because it represents the actual additional benefit of a coating. Anti-icing coatings on the rotor blades of wind turbines which, for example, lose their effectiveness after just a few weeks are clearly unsuitable for this technical application. There are test regions in Switzerland where icing phenomena on wind turbines are being studied (Gütsch, Switzerland, see Figure 3.29). There is also a special test stand on the Brocken (the highest peak in the Harz mountain range) for testing anti-icing coatings. The performance of a wide variety of coatings is being tested here under extreme weather conditions (see Figure 3.29).

3.2.4 Literature

[1] *Siegmann, K., Kaufmann, A., Hirayama, M.,* Anti-freeze Beschichtungen für Rotorblätter und Windenergieanlagen; Schlussbericht, 2006, source: *www.bfe.admin.ch/php/modules/enet/streamfile.php*

[2] *Atkins, P.W., De Paula, J.,* Atkins' Physical chemistry, 8th edition, Oxford University Press, 2006

[3] *Krzyzak, M., Techmer, K.S., Faria, S.H., Genov, G., Kuhs, W.F.,* Atomic Force Microscopy of Rearranging Ice Surfaces; In: *Kuhs, W.F.* (Edt.), Physics and Chemistry of Ice, RSC Publishing Cambridge, 2007, pp. 347-355

[4] *Faraday, M.,* Note on regelation; Proc. R. Soc., London, 10, 1860, 440-450

[5] *Rosenberg, R.,* Why Is Ice Slippery?, Physics Today; 2005, 50-55, source: *lptms.u-psud.fr/membres/trizac/Ens/L3FIP/Ice.pdf*

[6] *Jellinek, H.H.G.,* Liquid-like Layer on Ice; Journal of Colloid and Interface Science, 25, 1967, 192-205

[7] *Döppenschmidt, A., Butt, H.-J.,* Measuring the Thickness of the Liquid-like Layer on Ice Surfaces with Atomic Force Microscopy; Langmuir, 16, 2000, 6709-6714

[8] *Archer, P., Gupta, V.,* Measurement and Control of Ice Adhesion to Aluminium 6061 Alloy; J. Mech. Phys. Solids, 46, 10, 1998, 1745-1771

[9] *Atkins, P.W., De Paula, J.,* Atkins' Physical chemistry, 8th edition, Oxford University Press, 2006

[10] *Grunwald, I., Rischka, K., Kast, A.M., Scheibel, T., Bargel, H.*, Mimicking biopolymers on a molecular scale: nano(bio)technology based on engineered proteins, Phil. Trans. R. Soc. A, 367, 2009, 1727-1747
[11] *Ramlov, H.*, Aspects of natural cold tolerance in ecothermic animals, Hum Reprod 15 Suppl 5, 2000, 26-46
[12] *Duman, J.G., Bennett, V., Sformo, T., Hochstrasser, R., Barnes, B.M.*, Antifreeze proteins in Alaskan insects and spiders, Journal of Insect Physiology, 50, 2004, 259-266
[13] *Barrett, J.*, Thermal hysteresis proteins, Int J Biochem Cell Biol., 33 (2), 2001, 105-117
[14] *DeVries, A.L.*, Antifreeze Peptides and Glycopeptides in Cold-Water Fishes, Ann. Rev. Physiol., 45, 1983, 245-260
[15] Icing studies performed by Fraunhofer IFAM, Bremen, Germany (unpublished)
[16] *Anderson, D.N., Reich, A.D.*, Tests of the Performance of Coatings for Low Ice Adhesion, NASA Technical Memorandum, 1997, source: http://gltrs.grc.nasa.gov/reports/1997/TM-107399.pdf
[17] *Laforte, C., Laforte, J.-L., Carrière, J.-C.*, How a solid coating can reduce the adhesion of ice on structure; Contribution to IWAIS 2002, source: http://www.uqac.ca/amil/en/publications/papers/Icephobic%20 coatingsIWAIS2002.pdf
[18] *Zwieg, T.*, Ice Nucleating Non-stick Coating; Patent: WO 02/090459 A1; 2002
[19] *Laforte, C., Beisswenger, A.*, Icephobic Material Centrifuge Adhesion test, Contribution to IWAIS 2002, Source: http://www.uqac.ca/amil/fr/documentation/articles/IW53-CAT.pdf
[20] *Croutch, V.K., Hartley, R.A.*, Adhesion of Ice to Coatings and the Performance of Ice Release Coatings, Journals of Coatings Technology, 64 (815), 1992, 41-53

3.3 Self-healing coatings

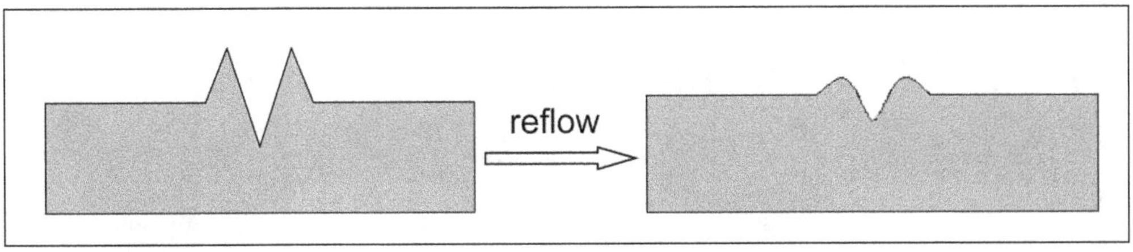

Figure 3.30: Reflow effect

Self-healing refers to the ability of materials to repair damage automatically without any additional external measures. Other terms for this behaviour are "self-repairing", "self-recovering", and "autonomic repairing".

In recent years a huge number of reports about self-healing materials have appeared. In this sense the term "self-healing" can be considered to be a buzz-word. The definition of "self-healing" coatings is a difficult matter. A lot of traditional coatings, for instance the classic chromate-containing anti-corrosion primers, can be considered to be self-healing materials because in these systems the chromate acts as a self-healing agent: In the event of damage, chromate crystals come into contact with the electrolyte, the chromate ions get solved and react with the metal substrate or with the oxide layer respectively, and so "heal" the damage in the coating. It is obvious that a book like this is unable to deal with all systems that can be considered to be "self-healing" in this sense.

As a consequence, the aspects of self-healing described here are limited to approaches based on recent technical developments and that deal with protective coatings. In this sense, protective coatings are corrosion-protection systems or coatings that can heal cracks or ruptures which would otherwise result in damage.

A short excursus will be given first of all though about coatings designed for aesthetic self-healing.

3.3.1 Aesthetic self-healing

The visibility of defects in a coating depends on the shape of the defects. The depth of such defects in a paint surface is usually not more than 0.5 µm [1]. If the edges of a defect are sharp, the amount of light that is reflected or scattered is high. Fresh defects are often sharp, and hence clearly visible. If the edges of the defect are slightly rounded, light cannot be reflected effectively and the defect is hardly, if at all, visible. Automotive coatings, for instance, should of course be as resistant to damage as possible. Making the coating harder to prevent the object causing the damage from penetrating the surface is not the appropriate way to achieve this. This is because the particles that lead to natural damage, for example sand grains or mineral dust, are normally harder than any clear coat which could be manufactured. The more elegant way of making coatings resistant to damage is to rely on an effect called "reflow" (Figure 3.30).

Reflow occurs when the temperature of the surface is above the glass transition temperature (T_g) of the coating. The surface tension together with the mobility of the polymers above the T_g lead to this smoothening effect.

A variety of coatings possessing this behaviour are available in the marketplace. The effect that can be achieved with this type of coating is shown in Figure 3.31.

Figure 3.31: Left: hood with fresh damage; Right: hood after reflow source: Nissan press release

Possessing resistance to damage is critical not only for automotive coatings, but also for coatings on floors, coatings for transparent plastics, and many other applications.

Aesthetic self-healing coatings and appropriate raw-materials are widely available in the marketplace and the technical approaches have been described in detail. The situation is very different for structural self-healing materials.

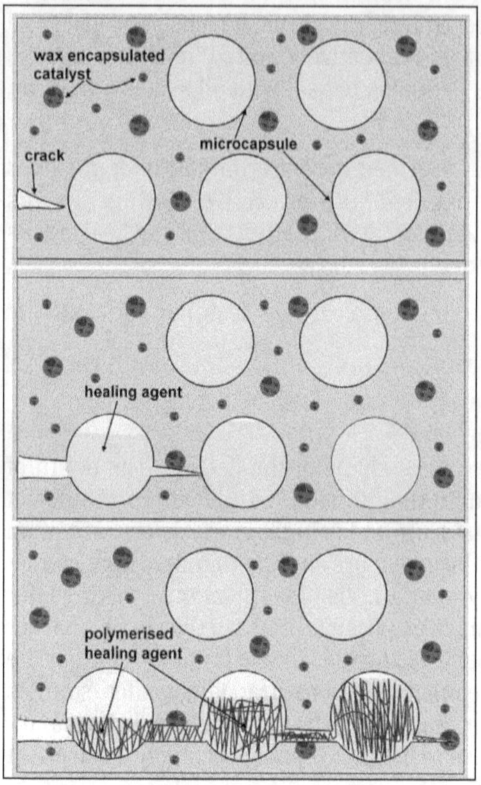

Figure 3.32: Top: Microcapsules filled with polymerisable healing agent and catalyst embedded in a polymer matrix; Middle: Crack ruptures capsules, healing agent leaks out, and comes in contact with catalyst; Bottom: Healing agent polymerises, seals the crack, and prevents crack propagation.

3.3.2 Structural self-healing

In the context of this book, the phrase "structural self-healing coatings" refers to coatings that are used to extend the lifetime of products which are exposed to high loads/stresses or to extend the inspection and repair intervals. Consequently, high corrosion protection, outdoor wood protection, and protection of reinforced plastic materials (e.g. rotor blades on wind turbines) are the focus of this section.

3.3.2.1 Microencapsulation of crosslinkable healing agents

The method of microencapsulation is widely used in industry for many different applications (see Chapter 2.2 "Microcapsules"). The first report on the use of microcapsules for creating self-healing polymeric materials was published in 2001 by White et.al. [2]. The idea behind this is illustrated in Figure 3.32.

The healing agent in this first example was dicyclopentadiene (DCPD) which can be effectively polymerised in a ring-opening metathesis reaction catalysed by Grubb's

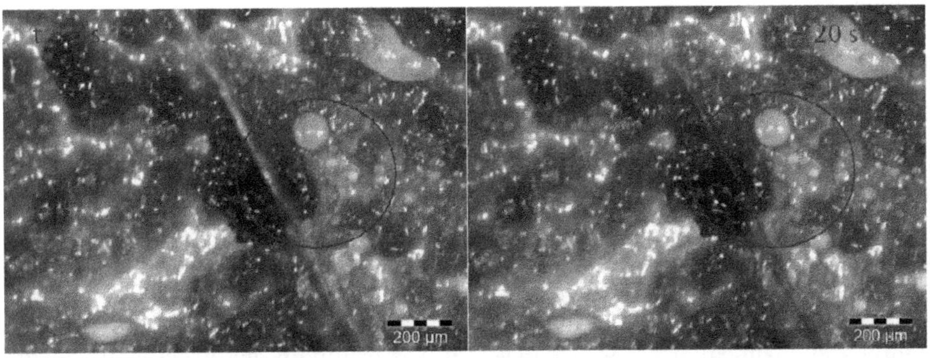

Figure 3.33: Microscope images of a model paint containing microcapsules with a crosslinkable blue resin. Left: freshly cut surface; Right: after 20 s the substrate material at the base of the cut is almost completely covered with the resin source: Fraunhofer IFAM[5]

catalyst, a ruthenium complex. It was shown that the toughness of an epoxy-based composite test sample could be recovered by this healing reaction and the healing efficiency was quantified [3]. The capsule sizes can range from approximately 100 µm down to 200 nm.

This chemical approach has also been evaluated for healing cracks and cuts in polyurethane based protective coatings [4]. For this purpose, a model paint formulation consisting of a two-pack aliphatic polyurethane was equipped with microcapsules containing DCPD and microcapsules containing Grubb's catalyst. DCPD was microencapsulated in a formaldehyde-urea polymer (see Chapter 2.2), and Grubb's catalyst was encapsulated in wax. It was demonstrated that cuts in such a system can be effectively "healed". Figure 3.33 shows the "healing" of a cut in a coating on a steel sheet. In a couple of seconds the blank steel at the bottom of the cut is covered with the DCPD (carrying a blue dye in this example to make the process visible) which subsequently polymerises. The capsule sizes are between 20 and 200 µm. The thickness of the capsule walls is approximately 100 to 150 nm.

The investigations on this chemical mechanism lead to two conclusions:

1. The concept of using microcapsules filled with polymerisable healing agents is feasible for protective coatings
2. The DCPD/Grubb's catalyst system is not practical because of the disgusting smell of DCPD and the limited durability and high price of Grubb`s catalyst

In recent years a variety of other microencapsulated and nanoencapsulated polymerisable systems have been developed which can be used for introducing a self-healing function into protective coatings:

- Methyl methacrylate together with catalysts [6]
- Bisphenol-A epoxy resin combined with a latent curing agent ($CuBr_2(2\text{-}MeIm)_4$, the complex of $CuBr_2$ and 2-methylimidazole) which is dispersed in the polymer matrix surrounding the microcapsules [7]. The healing reaction is very effective in terms of recovery of mechanical toughness of the polymer. The crosslinking occurs at temperatures above 120 °C so this approach is only feasible for applications where high temperatures can be applied to the damaged coating film [8].
- Linseed oil as a healing agent that is cured by ambient oxygen [9]

- Isophorone diisocyanate (IPDI) as a catalyst-free healing agent that cures via the ambient humidity [10]
- Epoxy resins [11] to be crosslinked with polythiol microencapsulated in the same way [12]. In this case the healing system is a two-part system, where both microencapsulated agents have to be used in optimal concentrations in the paint formulation.
- Polydimethylsiloxane (PDMS) based materials [13]. A system with very high chemical stability can be created using PDMS-based healing agents. One approach is based on hydroxy end-functionalized polydimethylsiloxane (HOPDMS) combined with polydiethoxysiloxane (PDES). The siloxanes are not miscible with the organic polymer matrix and are phase-separated in distributed microdrops. A catalyst, di-n-butyltin dilaurate (DBTL), starts the crosslinking reaction when cracks open the microdrops and the microcapsules containing the catalyst. The advantage of this system is the stability against environmental and chemical stresses. This system was originally not specifically designed for use in coatings. Most paint chemists would probably have concerns using PDMS in organic paint formulations. Meanwhile, microencapsulated PDMS-based healing agents designed for use in thermosetting coatings, elastomeric coatings, and powder coatings are commercially available and should be compatible with paint formulations, according to the information given by the manufacturers [14].

3.3.2.2 Self-healing anti-corrosion coatings with active agents

Although "functional coatings" in this book refer to coatings having functions which go beyond decoration and corrosion protection, self-healing anti-corrosion coatings will be dealt with at this point. This is appropriate because new developments are taking place in this area that are allowing the performance of conventional corrosion protection coatings to be significantly improved, and are allowing the replacement of very effective but unfortunately very toxic components such as chromium VI compounds.

The above-described systems are also suitable for prolonging the service life of coating systems for heavy corrosion protection. These systems, however, only provide passive corrosion protection due to the barrier effect of the healing resins, and the sealing of cracks or points of damage. Active corrosion inhibitors can also be incorporated into coating systems in an encapsulated form. These have long-term stability and are protected against leaching. A further advantage of using corrosion inhibitors in encapsulated form is that this gives greater freedom to the choice of inhibitors which can be used. In classic systems, inhibitors had to be chosen such that when an electrolyte solution came into contact with a point of damage sufficient inhibitor was released to hinder corrosive attack. On the other hand, the solubility in water had not to be so large that there were osmotic effects such as blistering. In addition, sufficient inhibitor has to be present for the entire service life of the coating. These boundary conditions considerably limited the choice of corrosion inhibitors, and particularly so for water-based coatings. During the search for alternatives to chromium VI compounds this was often a major obstacle. For the corrosion protection of aluminium alloys in the aviation sector (e.g. AA2024) that have a high copper content and are hence very susceptible to corrosion, the classic systems contain a high concentration of chromate compounds. With these systems, the ability to select from a variety of chromates having different solubility products (strontium chromate, calcium chromate, barium chromate) enabled the short-term effectiveness and also the long-term effect to be very effectively adjusted. As far as effectiveness is concerned, a number of chromate-free inhibitors are available. For example, benzotriazole and some benzotriazole derivatives provide very effective corrosion protection for copper-rich aluminium alloys. The problem here is the high solubility in water which causes osmotic effects and rapid exhaustion of the reservoir of inhibitor.

One way of solving this problem is to use encapsulated inhibitors or inhibitors affixed to support materials. The release of encapsulated inhibitors in the coating film can take place in various ways:

- customisable diffusion out of the support materials or capsules
- release triggered by pH changes which are initiated by the corrosion processes (see below)
- via mechanical damage to the capsules

One option for utilising corrosion inhibitors in encapsulated form is to use impregnated nanoparticles [15]. These particles can consist, for example, of a SiO_2 core covered by a corrosion inhibitor (e.g. benzotriazole). These particles are encapsulated using a layer-by-layer (LbL) process with an outer shell which protects the inhibitor and which is compatible with the coating matrix. The layer-by-layer process is based on the deposition of a sequence of layers of oppositely charged substances (polyelectrolytes, nanoparticles, biomaterials). Such materials have been tested for various applications, for example as supports for biocatalysts [16], as membranes for fuel cells [17], or for use as outer coating materials for nanoreactors [18]. For the present application, polyelectrolyte layers are used which are permeable to ions and small molecules but which hold back larger molecules. Such layers can be customised so that the release of corrosion inhibitors is, for example, triggered by pH or moisture. As corrosion processes go hand in hand with local changes in the pH of an electrolyte at the local anode or cathode [19], the release behaviour of corrosion inhibitors can be very elegantly controlled by this mechanism (see Figure 3.34). The benefit of nanoparticles structured in this way has, for example, been demonstrated [15] for the protection of copper-rich and hence very corrosion-susceptible aluminium alloys such as the alloy AA2024 that is commonly used in the aviation industry.

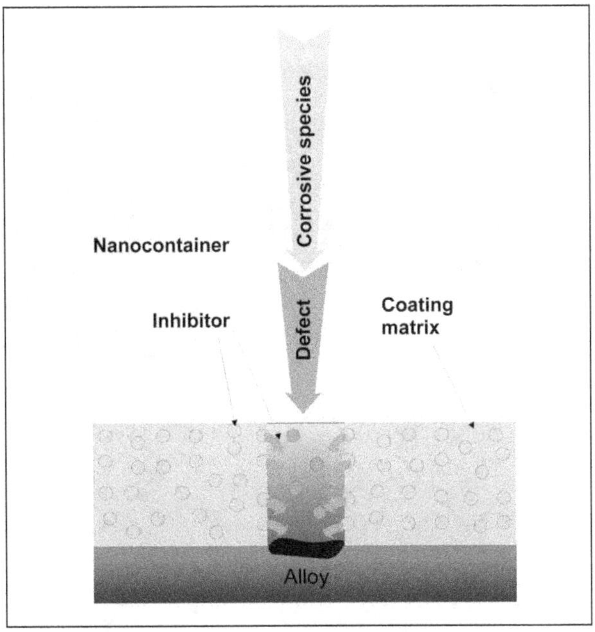

Figure 3.34: Corrosion protection mechanism using nanoparticles with affixed corrosion inhibitor [15]
source: H. Möhwald, MPI of Colloids and Interfaces

Organic inhibitors can also be used in a microencapsulated form in epoxy primers. Such systems have been intensely studied for the corrosion protection of steel. For this purpose, sand-blasted steel samples were coated with an EP primer (containing inhibitor-filled microcapsules, and also without microcapsules for reference purposes). The samples were then given a polyurethane topcoat. Both electrochemical analyses (electrochemical impedance spectroscopy, EIS) and storage under laboratory and natural conditions demonstrated significant improvement in the corrosion protection [20]. Unfortunately, the authors provided no information whatsoever about the nature of the inhibitor.

3.3.2.3 *Other approaches for self-healing coatings*

In recent years a series of other approaches have been described which could lead to the development of industrially viable, self-healing coatings. A few of these approaches are described below.

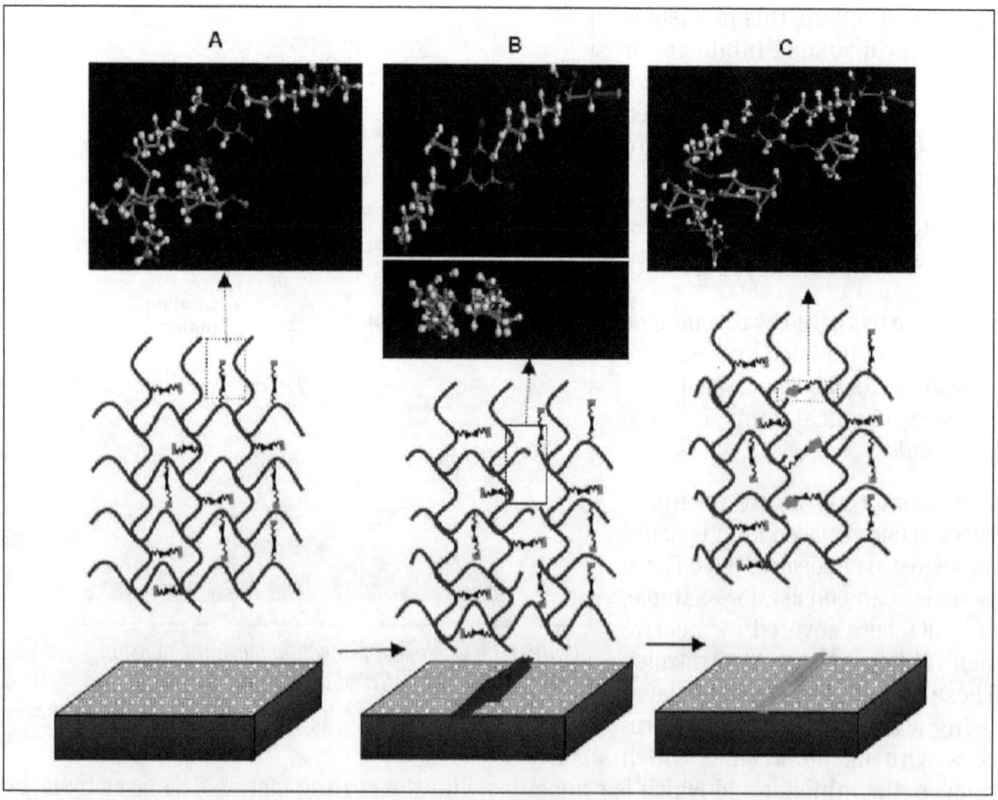

Figure 3.35: Repair mechanism for polyurethane modified with oxetane-substituted chitosan [21]
source: American Association for the Advancement of Science

Self-healing polyurethane via modification with oxetane-substituted chitosan [21]

Self-healing mechanisms do not necessarily have to be based on reservoirs of healing reagents. Some current studies are showing promise here. A good example is an approach using polyurethane coatings modified with oxetane-substituted chitosan. A model coating system was manufactured based on hexamethylene diisocyanate (HDI), polyethylene glycol (PEG), and an oxetane-substituted chitosan derivative. Chemical bonds, which were damaged by mechanical effects, were chemically repaired using this system. Figure 3.35 schematically shows this process. Status A in the figure shows the undamaged polymer network, status B represents the network damaged by mechanical loads, and status C depicts the "repair" of the damage via the relevant functional groups present in the network. The "repair" occurs via the formation of new chemical bonds with the help of the relevant functional groups which are activated by UV light.

This self-healing mechanism is evidently able, as demonstrated in Figure 3.36, to heal scratches or cracks of a few microns in width by irradiation with UV light. A scratch of almost 10 µm width could apparently be completely healed by irradiation for 30 minutes with UV light of 302 nm wavelength (120 W). The top part of Figure 3.36 shows infrared images which clearly show the heat tones of the chemical reactions.

Self-healing via a retro Diels-Alder reaction [23, 24]

Another approach for the development of self-healing materials is based on a reversible polymerisation reaction. In this case the polymer at the point of damage is first of all decomposed by heat into monomers or oligomers. After cooling, these are then reacted again to

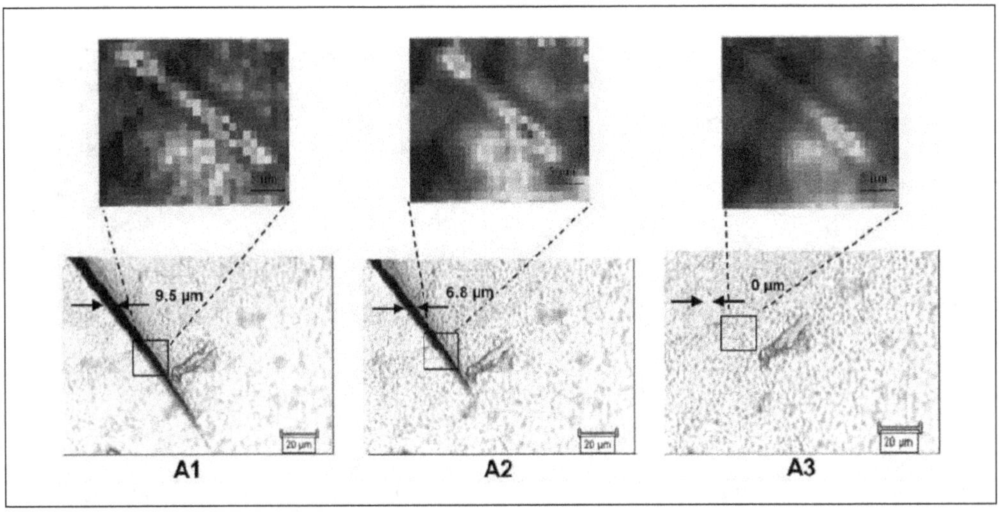

Figure 3.36: IR images (upper) and optical images (lower) of OXE-CHI-PUR networks at different UV exposure times: A1, 0 min; A2, 15 min; A3, 30 min [21, 22] source: Elsevier Ltd

form a polymer network. Compared to the previously described approach, this method of "healing" can theoretically be repeated as often as desired. One example of such a polymer system is a product formed by a thermally reversible Diels-Alder cycloaddition reaction from a multi-diene (multi-furan) and a multi-dienophile (multi-maleimide) [22]. It was demonstrated here that the polymer attained more than 60 % of its original strength after successful healing of the damaged site. The healing process could apparently also be followed visually (see Figure 3.37).

Successful healing results can be achieved using this approach, although the polymer is probably less stable to weathering and somewhat yellow due to the large number of double bonds. This would probably considerably limit the practical use of this approach.

Self-healing polymer materials by embedding filled hollow fibres [26]

In the area of fibre reinforced composite materials, development work has been ongoing for a number of years to give the materials self-healing properties and to make any damage visible. These properties would have enormous benefits when such materials were used for aircraft and car manufacture and for wind turbines. One approach here involves the use of hollow glass fibres that are filled with healing reagents and also with fluorescent dyes. The

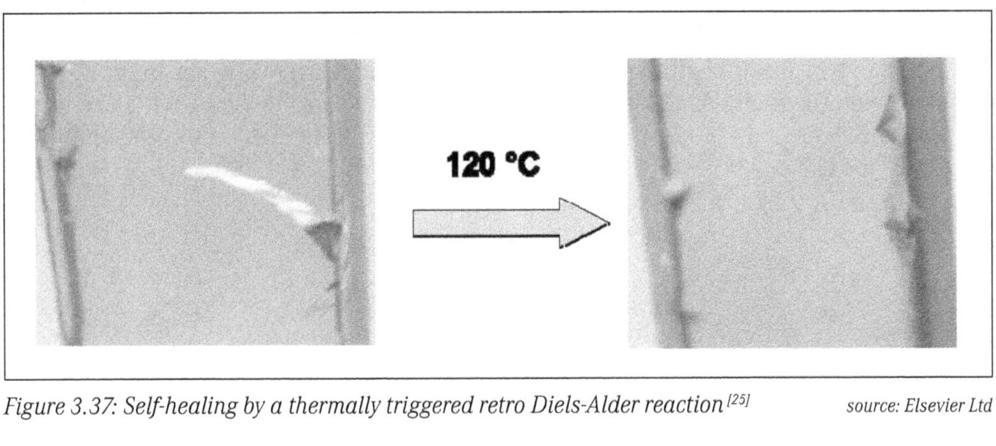

Figure 3.37: Self-healing by a thermally triggered retro Diels-Alder reaction [25] source: Elsevier Ltd

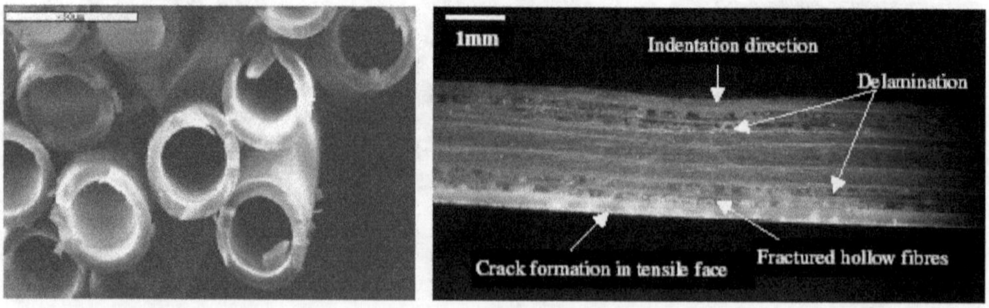

Figure 3.38: Left: Hollow fibres; Right: Healing a coating, made visible by fluorescent dye [26]

source: Elsevier Ltd

latter make points of damage visible under UV light and allow the healing of the material to be visualized [26]. This application uses hollow glass fibres of diameter 60 µm and a hollow volume of 50 % in an epoxide matrix. The fibres contain the fluorescent dye, and half are filled with epoxy resin and the other half with the relevant hardener. If an external effect causes the fibres to fracture, the components are released and react together to form a polymer. This partly recovers the strength of the material and hinders propagation of the damage (see Figure 3.38).

This approach was originally intended for composite materials, but it is conceivable to also use this method for protective coatings which are subject to extreme loads (e.g. offshore structures).

3.3.3 Industrial application and outlook

The idea of developing self-healing materials still sounds more like science fiction than industrial reality. The approaches described here are all, to a greater or lesser extent, still at a conceptual stage. Industrial implementation is still some years away. By far the most advanced approaches are those based on the use of reservoirs filled with active healing agents. In the next few years it is expected that the first commercial suppliers will have relevant raw materials and coatings available [14]. A new impulse for the development of self-healing coatings, which should prolong the lifetime of capital equipment, will certainly come from the establishment of new technologies, for example offshore wind energy. In this case, the profitability of a wind farm is highly dependent on the service life of the components. This is because inspection, maintenance, and any necessary repairs to surface protection are, if possible at all, extremely expensive.

The extent to which self-healing materials will break through into a wider range of areas will depend on the fundamental attitude of industry to the subject of longevity. The fact is that not every company in our throw-away society is really interested in prolonging the lifetime of goods.

3.3.4 Literature

[1] *Wicks, Z.W.Jr.; Jones, F.N.; Pappas, S.P.; Wicks, D.A.,* Organic Coatings, Science and Technology, Third Edition, Wiley-Interscience, Weinheim, 2007, p 85

[2] *White, S.R.; Sottos, N.R.; Geubelle, P.H.; Moore, J.S.; Kessler, M.R.; Sriram, S.R.; Brown, E.N.; Viswanathan, S.* Autonomic healing of polymer composites, Nature, 409, 2001, 794-797

[3] *Blaiszik, B.J.; Sottos, N.R.; White, S.R.*, Nanocapsules for self-healing materials, Composites Science and Technology 68, 2008, 978-986

[4] *Stenzel, V.; Mock U.; Tillner, S.* DIY repairs - self-healing mechanisms are beeing developed for protective coatings, European Coatings Journal, 11, 2006, 32-36

[5] *Mock U., Tillner, S., Stenzel, V.*, Self-healing concepts for protective coatings, Proceedings Nuremberg Congress 2007

[6] *He, X.; Shi, X*, Self-repairing coating for corrosion protection of aluminium alloys, Progress in Organic Coatings 65, 2009, 37-43

[7] *Yin, T.; Rong, M.Z.; Zhang, M.Q., Yang, G.C.*, Self-healing epoxy composites - Preparation and effect of the healant consisting of microencapsulated epoxy and latent curing agent, Composites Science and Technology 67, 2007, 201-212

[8] *Gosh, S.K.*, Self-healing materials - fundamentals, design strategies and applications, Wiley-VCH, 2009, page 42

[9] *Suryanarayana, C.; Rao, K.C.; Kumar, D.*, Preparation and characterisation of microcapsules containing linseed oil and its use in self-healing coatings, Progress in Organic Coatings, 63, 2008, 72-78

[10] *Yang, J.; Keller, M.W.; Moore, J.S.; White, S.R.; Sottos, N.R.*, Microencapsulation of isocyanates for self-healing polymers, Macromolecules, 41, 2008, 9650-9655

[11] *Xao, D.S.; Rong, M.Z.; Zhang, M.Q.*, A novel method for preparing epoxy-containing microcapsules via UV irradiation-induced interfacial copolymerization in emulsions, Polymer, 48, 2007, 4765-4776

[12] *Yuan, Y,C,; Rong, M.Z.; Zhang, M.Q.*, Preparation and characterization of microencapsulated polythiol, Polymer, 49, 2008, 2531-2541

[13] *Cho, H.S; Andersson, H.M.; White, S.R.; Sottos, N.R.; Braun, P.V.*, Polydimethylsiloxane-based self-healing materials, Advanced Materials, 18, 2006, 997-1000

[14] www.autonomicmaterials.com

[15] *Zheludkevich, M.L., Shchukin, D.G.; Yasakau, K.A.; Möhwald, H.; Ferreira, M.G.S.*, Anticorrosion coatings with self-healing effect based on nanocontainers impregnated with corrosion inhibitor, Chem.Mater. 19, 2007, 402-411

[16] *Yu, A.; Liang, Z.; Caruso, F.*, Enzyme multilayer porous membranes as biocatalysts, Chem. Mater., 17, 2005, 171-175

[17] *Farhat, T.R.; Hammond, P.T.*, Designing a new generation of proton-exchange membranes using a layer-by-layer deposition of polyelectrolytes, Adv. Funct. Mater. 15, 2005, 945-954

[18] *Shchukin, D.G.; Sukhorukov, G.B., Möhwald, H.*, Biomimetic fabrication of nanoengineered hydroxyapatite/polyelectrolyte composite shell, Chem. Mater., 15, 2003, 3947-3950

[19] *Yasakau, K.A.; Zheludkevich, M.L.; Sviatlana, S.V., Ferreira, M.G.S.*, Mechanism of corrosion inhibition of AA2024 by rare-earth compounds, J. Phys. Chem. B, 110 (11) 2006, 5515-5528

[20] *Metha, N.K.; Bogere M.N.*; Environmental studies of smart/self-healing coating system for steel, Progress on Organic Coatings, 64, 2009, 419-428

[21] *Gosh, B.; Urban, M.W.*, Self-repairing oxetane-substituted chitosan polyurethane networks, Science 232, 2009, 1458-1460

[22] *Urban, M.W.*; Stratification, stimuli responsiveness, self-healing, and signalling in polymer networks, Progress in polymer science 34, 2009, 679-687

[23] *Chen,X.,: Matheus A. Dam, M.A.; Ono, K.; Mal, A.; Shen, H.; Nutt, S.R.; Sheran, Wudl, F.*, A thermally re-mendable cross-linked polymeric material, Science 265, 2002, 1698-1702

[24] *Chen, X; Wudl, F; Ajit K. Mal, A.K.; Shen, H.; Nutt, S.R.*, New Thermally Remendable Highly Cross-Linked Polymeric Materials, Macromolecules 36, 2003, 1802-1807

[25] *Gould, P.* Self-help for ailing structures, Materials Today, June 2003, 44-49

[26] *Pang, J.W.C; Bond, I.P.*, A hollow fibre reinforced polymer composite encompassing self-healing and enhanced damage visibility, Composites Science and Technology 65, 2005, 1791-1799

3.4 Drag-reducing surfaces

In autumn 2009 the European Union decided that CO_2 emissions from the aviation industry and shipping industry must be reduced by 10 and 20 % respectively, with corresponding implications for fuel consumption. In the future the CO_2 emission of aircraft and ships will be measured, and financial levies (Emission Trading System) will be calculated.

These boundary conditions, and also the envisaged rising cost of fuel in the future, mean that all technical means of reducing fuel consumption must be utilised. Surface technology can also make a valuable contribution here because the latest technology allows the drag of surfaces to be significantly decreased by applying suitable coating systems.

3.4.1 Laminar and turbulent flow

The drag of cars, aircraft, ships, rail vehicles, and also of rotor blades on wind turbines comprises several components:

- Pressure drag due to the pressure difference between the side experiencing the flow (e.g. the bow of a ship) and the opposite side (e.g. the stern)
- Induced drag on objects that generate lift (e.g. aircraft wings)
- Wave drag, for example on ships due to the bow wave and stern wave
- Frictional drag due to the friction of the fluid (air or water) at the surface

The frictional drag, namely the friction between the solid (coating) surface and the flowing fluid, can be influenced considerably by the properties of the coating. As the frictional drag plays a major role for ships and aircraft in particular (for ships it is sometimes more than half the total drag [1]), it is worth considering the options available for influencing the frictional drag by means of coating technology.

The frictional drag is highly dependent on the flow conditions. In the 19th century the dimensionless Reynolds number was introduced to describe the flow conditions:

Equation 3.1 $$Re = \frac{v \cdot l \cdot \rho}{\mu}$$

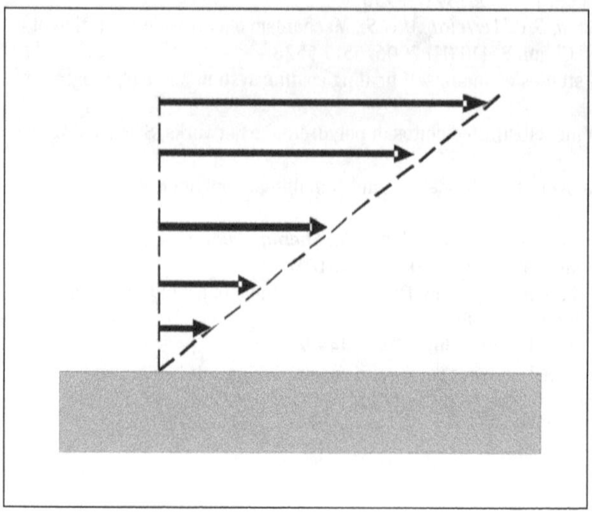

Figure 3.39: Velocity profile for laminar flow (length of arrows indicates the velocity). The flow velocity is zero at the solid surface.

where v = flow velocity,
l = characteristic length,
ρ = density of the fluid, and
μ = dynamic viscosity of the fluid.

If Re is small, the flow is laminar. The flow is free of vortices and turbulence, resulting in minimum frictional drag (which depends on the viscosity of the fluid). The typical flow profile is shown in Figure 3.39.

If the flow is laminar, the frictional drag of the fluid at the surface is small. If the flow velocity and/or the length increases, the flow changes to turbulent flow after going through an unstable transition (see Figure 3.40).

Waves develop in the transition region (Tollmien-Schlichting waves) which then become larger and eventually change into wholly turbulent flow. Vortices develop in the turbulent region. These vortices have components in both the flow direction (streamwise) and perpendicular to this (spanwise) (see Figure 3.41 and 3.42). The Reynolds number at which this transition occurs (namely at what velocity and length) depends on a variety of factors including the surface structure of the wall and has to be determined for each individual case. This transition is, for example, at ca. 2300 for tubular flow.

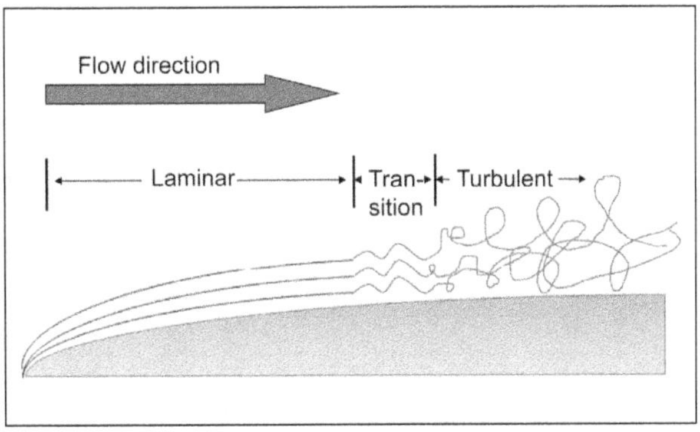

Figure 3.40: Transition of laminar flow into a turbulent boundary layer

Even minor interference (e.g. contamination of the surface by insects on the rotor blades of a wind turbine) can cause laminar flow to switch to turbulent flow. The drag associated with turbulent flow is considerably greater than with laminar flow. It follows that one endeavours to maintain laminar flow for as long as one can. When this is no longer possible, it is endeavoured to reduce the turbulent wall friction using a suitable method. What options are currently available for influencing the frictional drag using a coating surface?

3.4.2 Riblet surfaces - artificial sharkskin structures

It has been known for many years that the skins of fast-swimming sharks have a riblet structure in the length direction (see Figures in Chapter 2.1). The knowledge that these sharks have such an outer surface made researchers realise that this riblet structure on the scales has a fluid mechanical benefit. Indeed, this feature has continued to evolve over the past 100 million years or so. Detailed studies at the Deutsches Zentrum für Luft- und Raumfahrt (DLR) [2, 3] and other research establishments [4] have elucidated the mechanism by which the wall friction is reduced and what reduction in wall friction can be expected from technical surfaces that function using the same principle.

At first thought it might be expected that a perfectly smooth surface would have the lowest wall friction. Under certain flow conditions, however, surfaces with riblets specially adapted to the flow conditions have a distinct advantage. The effect of riblet-structured surfaces is only evident when there is turbulent flow.

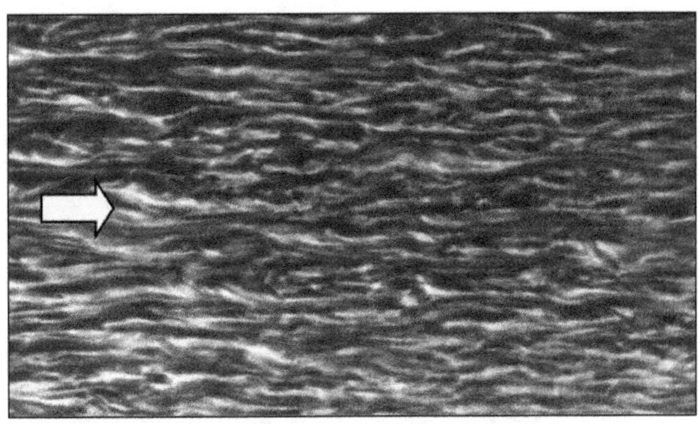

Figure 3.41: Flow visualisation in water with aluminium powder close to the wall - the arrow indicates the flow direction (Van Dyke, M.). The flow field in all instances of turbulent flow close to the wall is dominated by streak structures, which are counter-rotating vortices.

Source: Cambridge University Press

Turbulent flow, similar to laminar flow, has a main flow direction. For turbulent flow, however, this is also superimposed by oscillations transverse to the main flow. In the direct vicinity of the wall streak-

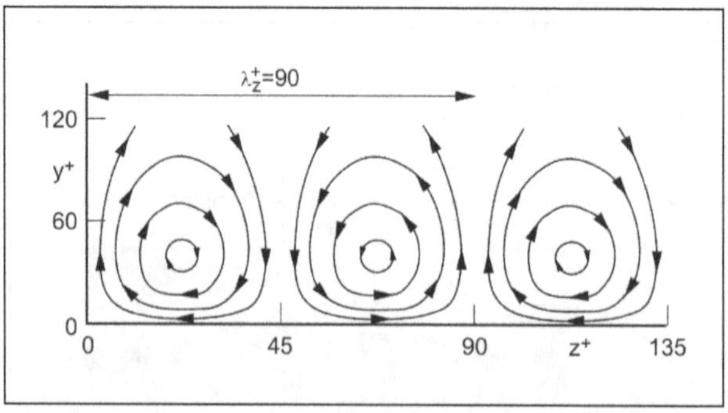

Figure 3.42: Turbulent transverse flow perpendicular to the main flow (direction of main flow perpendicular to paper surface) [5] *source: Cambridge University Press*

like structures form which can be thought of as counter-rotating vortices. These structures are shown in Figure 3.41. Figure 3.42 shows an (idealised) section through these structures transverse to the flow direction.

These oscillations, in particular those transverse to the main flow direction (see Figure 3.42), are the cause of the significantly higher wall friction in turbulent flow.

The riblets parallel to the main flow direction cause the origins of the longitudinal and transverse components of the turbulent flow to be differently affected (Figure 3.43). The transverse component, which is primarily responsible for the higher turbulent drag, is affected more and its origin is further from the wall than that in the main flow direction. This difference means that less friction is transferred to the wall.

Somewhat more graphic, though not wholly correct, is the explanation that the vortex components in the transverse direction now only act at the crests of the riblets and contribute less to the frictional drag than for smooth surfaces.

In experiments with adjustable riblet surfaces in an oil tunnel it was found that the ideal height of the riblets is half the distance apart [7]. The effective riblet spacing depends in turn on the flow conditions at the surface, as defined by the Reynolds number (Re). The higher the Reynolds number the smaller the distance apart and height of optimal riblets. The distance apart and height are in the µm range for many technically relevant flow conditions. For an 8 m long object in water and a flow velocity of 10 m/s, a riblet interval of ca. 50 µm is optimal (corresponds to a height of 25 µm). For aircraft this value typically lies between 50 and 100 µm depending on the velocity, size, and altitude.

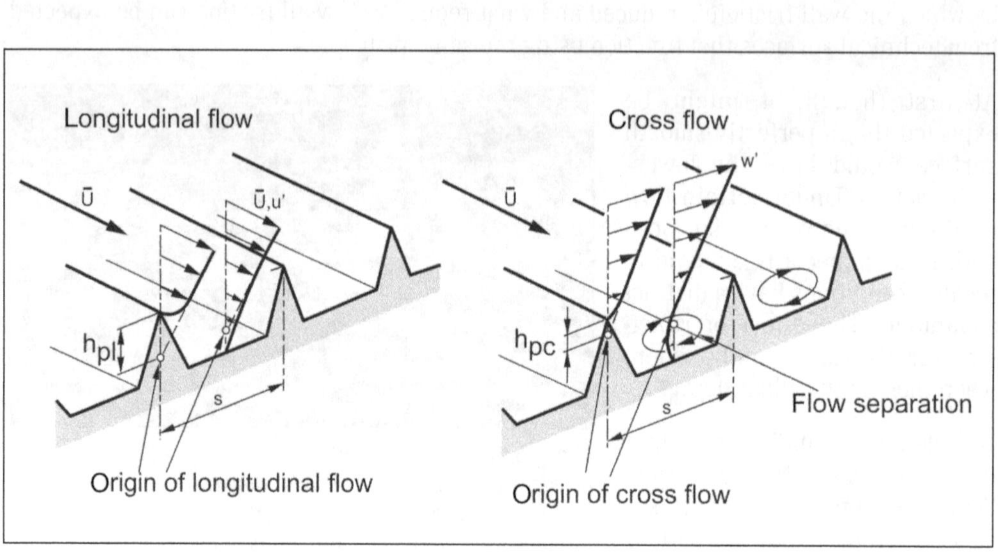

Figure 3.43: Effect of riblet-grooves on the components of turbulence perpendicular to the main flow
source: Wolfram Hage, DLR [6]

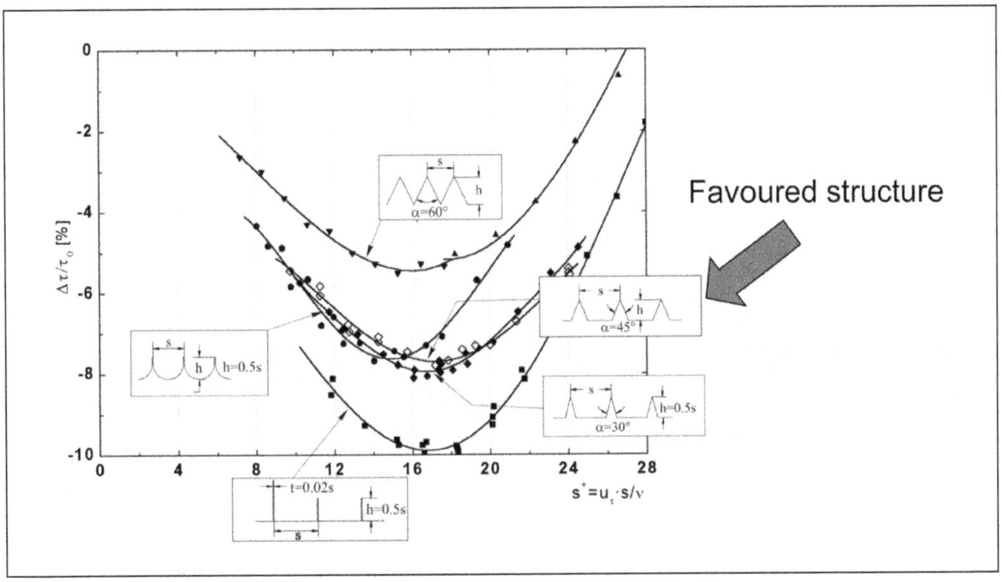

Figure 3.44: Effect of different riblet geometries on drag reduction. Vertical axis: drag reduction (%), horizontal axis: dimensionless value indicating flow condition (comparable to a Reynolds number),
source: Wolfram Hage, DLR

The shape of the riblets also has a decisive effect on the drag reduction. Figure 3.44 shows the effect of cross-section geometry on riblet effectiveness. The percentage reduction in wall drag is shown on the vertical axis ($\Delta\tau/\tau_0$). The horizontal axis shows a value which can be considered to be the dimensionless velocity or riblet interval and corresponds to the Reynolds number. Ideally thin lamellae in the flow direction have the largest effect (reduction in wall drag of 10 %). However, such lamellae cannot be produced using coatings and would not be very stable from a mechanical point of view. Simple saw-tooth structures give a reduction of about 5 %. Riblet geometries with a crest angle of 45° have been demonstrated to be the best compromise between effectiveness and manufacturability/stability. These can be manufactured with high quality using polymeric materials and give a reduction in wall drag of 7 to 8 %.

The best results were obtained for a riblet surface where the direction of the riblets does not differ by more than ± 10° from the main flow direction.

Since the 1980s and 1990s experiments have been conducted on riblet-structured films in various industries to demonstrate the potential of riblet surfaces. Examples include:

In February 1987 the American yacht “Stars and Stripes” won the America’s Cup. The yacht had a riblet surface to reduce the wall-related drag [8].

An Airbus A340 was partially covered with a riblet film. When in service with Cathay Pacific this resulted in a fuel saving of 1 to 1.5 %. Only part of the surface could be covered with the film, but based on the whole surface it is estimated that fuel savings of about 2 to 3 % are possible using this technology [9]. The film, however, adds to the weight of the aircraft which decreases the effective saving.

Although these experiments demonstrated potential applications for riblet technology, other disadvantages, besides the additional weight, became apparent:

A film cannot be applied crease-free to doubly curved surfaces (a cylinder can be covered crease-free with a film, but a sphere not).

Figure 3.45: Riblets in a dual-cure coating source: Fraunhofer IFAM

Film application is complex by hand.

Removing the film after several years in use is very difficult (aircraft are stripped of paint about every 5 years).

One way of overcoming the disadvantages of film detachment is to use a riblet-structured coating. Such a coating can, for example, be applied to large and also curved surfaces using the roller method described in Chapter 2.1.4.

As the structures at the micron level must remain intact over a long period, the coating used on aircraft and wind turbines must have extremely good mechanical and weathering properties. For example, very good results have been achieved using a solvent-free, nanoparticle reinforced dual-cure coating [10]. In such coating systems, the UV curing fraction is kept optimised to ensure that the coating does not reflow and is tack-free. The resulting riblet surfaces were successfully tested using laboratory methods (Amtec-Kister brush test and QUV weathering test) and also via in-service tests on test areas on an aircraft in commercial operation. Figure 3.45 shows a scanning electron micrograph of the riblets produced using this dual-cure system.

Furthermore, a wing section of 1.5 m width was coated with such a riblet coating and tested under suitable conditions in a wind tunnel by the Deutsches Zentrum für Luft- und Raumfahrt (DLR) in Berlin (Figure 3.46).

Compared to a similar section with a smooth and unstructured surface, the coated surface gave a reduction in the total drag of 6.25 %. This result shows that riblet surfaces are advantageous for aerodynamic sections (airfoils and rotor blades of wind turbines).

It is evident that a riblet surface also results in reduced drag when water rather than air is the flowing medium (the effect was indeed first discovered for sharks). Experiments on an 8 m long, torpedo-shaped object carried out by the Hamburgische Schiffbau-Versuchsanstalt (HSVA) have shown drag reduction of up to 5.2 %. This work compared a riblet-structured object to the same object having a smooth but chemically identical surface. This result indicates that the application of a riblet surface on a ship can give a fuel saving of about 2.5 %, assuming that the wall drag makes up about 50 % of the total drag. The biggest challenge here is to avoid biological fouling/growth, which would very rapidly destroy the desired effect. Reports that a riblet surface (or sharkskin surface) has anti-fouling properties merely due to the microstructure are however wishful thinking.

Figure 3.46: Wing section for drag measurement in a wind tunnel with 95 % of the chord covered with riblet coating source: Wolfram Hage, DLR

3.4.3 *Hydrophobic and superhydrophobic surfaces*

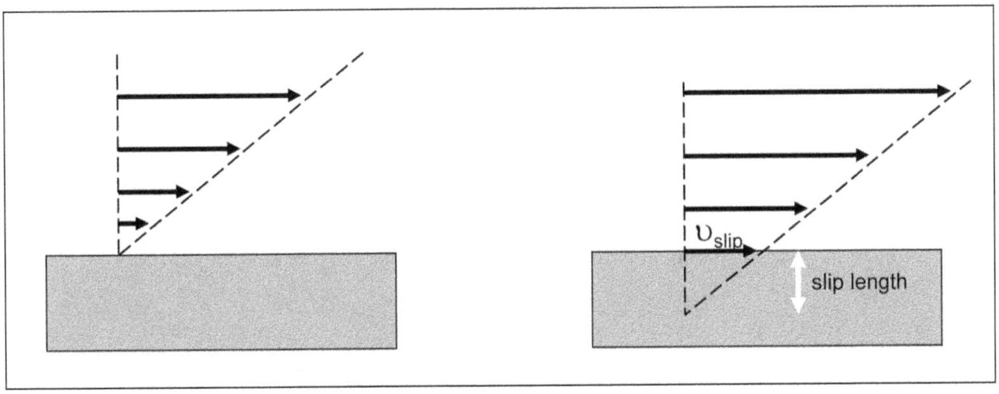

Figure 3.47: Left: non-slip boundary condition. Right: slip boundary condition and definition of slip length

The velocity of a viscous fluid (e.g. water or air) directly at an interface is normally considered to be 0 in fluid dynamics (see Figure 3.39) and this has been substantiated in numerous studies. This condition is also known as the no-slip condition.

Different behaviour is however found in certain cases for hydrophobic surfaces and in particular for superhydrophobic and ultrahydrophobic surfaces in water. In these cases, the velocity directly at the wall surface is not zero. This condition is known as the slip condition. A measure often used for this property is the slip length. This is measured by extrapolating the velocity behind the surface to zero and then determining the distance to the surface (see Figure 3.47). The larger the slip length, the greater the potential for reducing the drag.

In order to estimate whether these types of surfaces can be used to reduce the drag of an object, a number of cases must be distinguished:

- **Laminar flow**
  Laminar flow represents the optimum case for drag reduction using hydrophobic surfaces. If, however, the velocity directly at the surface of the object is not zero but rather a slip boundary condition prevails, the drag of the fluid can still be significantly reduced. On an ultrahydrophobic surface (organosilane modified, microstructured silicon wafer) whose wetting properties follow the Cassie regime (see Chapter 2.1.2.2), a flow velocity of more than 60 % the average velocity was, for example, measured directly at the surface [11]. This corresponds to a slip length of 7.5 μm. In other experiments on ultrahydrophobic surfaces a slip length of even 20 μm was measured. For laminar flow this led to drag reduction of 40 %[12].

For laminar flow, a slip boundary condition in all cases leads to drag reduction. For turbulent flow, which is considerably more important in technology and engineering, the situation is however different.

- **Turbulent flow**
  For turbulent flow, different cases must again be distinguished. This is because, as described above, there are vortex components in the main flow direction and also perpendicular to this (see Figure 3.43). Hydrophobic surfaces with a slip boundary condition show different behaviour depending on the direction in which they are effective [13]. From theoretical calculations [14] it follows that: If a surface only had a slip boundary condition in the flow direction and no slip behaviour perpendicular to this, the wall friction reduction would be a maximum and the effect optimal.

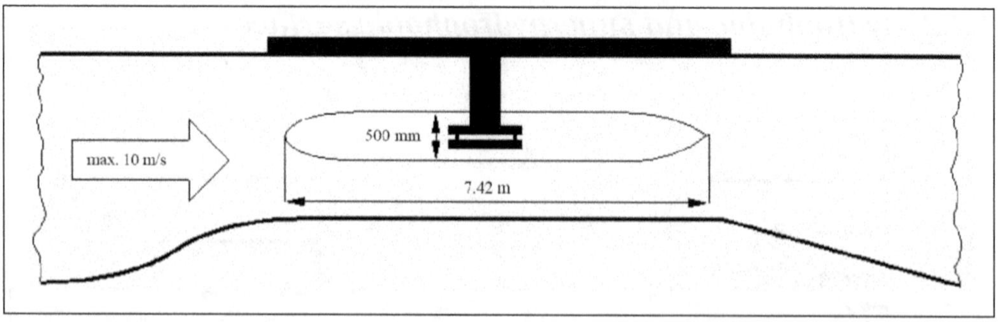

Figure 3.48: Device for measurement of drag under conditions relevant for shipping at HSVA, Hamburg/Germany

In contrast, if a surface only has a slip boundary condition perpendicular to the main flow direction, and no slip behaviour in the main flow direction, this leads to a significant increase in the wall friction and drag.

A surface which has slip boundary conditions in both directions, like a homogenous, hydrophobic surface in water, leads to a reduction in drag, but this is far less pronounced than for a surface which only has a slip boundary condition in the flow direction.

The above quoted work on drag on hydrophobic surfaces was either purely theoretical or was carried out on a micro-scale under laboratory conditions.

In order to study the effect of hydrophobic coatings in practically relevant tests, the authors of this book commissioned drag tests on 7.4 m long, torpedo-shaped objects in collaborative work with the Hamburgische Schiffbau-Versuchsanstalt (HSVA). These tests were carried out in a flow tunnel which permitted water velocities of up to 10 m/s and hence Reynolds numbers relatively close to those under real conditions on a ship's hull (see Figure 3.48).

A comparison was made between a classic self-polishing anti-fouling coating (reference, current state of technology) and three different hydrophobic coatings:

a: Commercial fouling-release silicone coating 1 (surface energy 22 mN/m)
b: Commercial fouling-release silicone coating 2 (surface energy 22 mN/m)
c: Hydrophobic polyurethane coating (surface energy 15 mN/m)

The results are shown in Figure 3.49.

Interestingly, each of the three hydrophobic coatings shows similar behaviour compared to the reference. The drag compared to the reference initially increases with increasing Reynolds number. Above a Reynolds number of ca. $2.5 * 10^7$ the drag becomes increasingly smaller. At high Reynolds numbers in the region in which the conditions come closest to those for a ship (for a 150 m long ship sailing at 15 knots the Reynolds numbers are in the region $9 * 10^8$ to $1 * 10^9$), the drag shown by all the hydrophobic coatings is a few percent lower than the drag for the classic anti-fouling coating.

Even though the authors do not fully understand this phenomenon, the results confirm that hydrophobic coatings offer an advantage with regards to drag in water. Further research work is required in order to draw more detailed conclusions.

It can be deduced from the aforementioned theoretical discussion and the experimental values that the combination of riblet coatings (see Figure 3.43) with a hydrophobic surface having a slip boundary condition in only the flow direction could provide an optimum solution. Experimental proof of this has not yet been obtained.

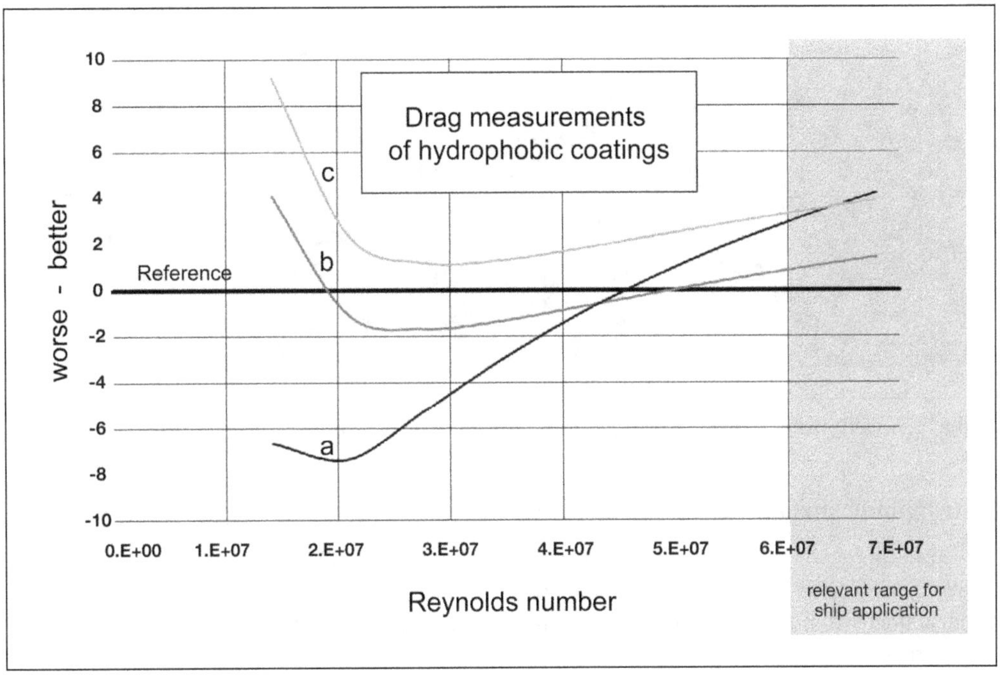

Figure 3.49: Drag measurements on hydrophobic coatings, performed by HSVA Hamburg/Germany, further explanation in the text above

3.4.4 Air-trapping surfaces

Superhydrophobic coatings which follow the Cassie regime (Chapter 2.1.2.2) have air inclusions between the wetting liquid and the solid surface. Air, whose viscosity is 55 times less than that of water, is a very effective lubricant at this interface and can lead to significant reduction in the wall friction. The fact that such surfaces have a slip-boundary condition is amongst other things due to this.

An obvious thought is to then use air as a lubricant to reduce fuel consumption by ships. In Japan this concept was studied in detail in the late 1990s. Equipment was fitted on a research ship (the Saiun-Maru) to bathe the hull with air bubbles when the ship was traveling[15]. This involved liberating up to 110 m^3 air per minute in the form of micro-bubbles. In total 50 test voyages were conducted at speeds of between 14 and 21 knots. The tests gave the following results:

- Micro-bubbles can markedly reduce the friction;
- the technique also works under rough conditions in the open sea;
- the energy required to continuously pump air into the water using compressors could under some circumstances be greater than the energy saving due to the reduction in wall friction.

The latter point highlights that this simple approach is not so promising. So what can be done to keep the air longer at the surface? This calls once again for the use of a functional coating system.

In the first instance as hydrophobic a coating as possible would appear favourable here because this would ensure that air and not water is preferentially at the surface. The air should if possible be held on the surface so that its lubricating effect remains for as long as possible. A structuring of the surface which allows air to be trapped and held in hydrophobic voids is necessary for this. The following approaches can be adopted for this:

Figure 3.50: Left: Water spider with air-covered surface. Right: Microscopic image of the hairy surface

source: Prof. Dr. Wilhelm Barthlott, University of Bonn/Germany

Hydrophobic surfaces with hairs/fibres

This approach is being actively pursued under the banner of bionic research. A water spider *(Anylometes bogotensis)* serves as a model here. This spider is surrounded by an air envelope underwater which is made possible by the water-repelling hairs on the body of the insect (see Figure 3.50).

Figure 3.51: Model boat with flock-coated surface

source: Prof. Dr. Wilhelm Barthlott, University of Bonn/Germany

Technically this can be mimicked, for example, by flock-coating a surface with hydrophobic fibres. In this case the result is, like with the spider, a glossy silvery air layer underwater as shown in Figure 3.51.

The authors own studies on drag reduction in collaboration with the Centre of Applied Space Technology and Microgravity (ZARM) have confirmed the reduction in the wall friction. In this work two plates of DIN A4 size were flock-coated. One of the plates was made hydrophobic by applying a plasma polymer coating (see Figure 3.52).

Figure 3.52: Water drops on a flock-coated surface after plasma polymeric hydrophobic treatment

source: Fraunhofer IFAM

The drag coefficient (cd) of both plates was determined in a water flow tunnel by measuring the velocity of the wake. The results demonstrated that the hydrophobic, flock-coated surface had a drag coefficient of 0.006, with the corresponding value for the non-hydrophobic surface being 0.011. The air layer arising from the hydrophobic treatment therefore virtually halved the wall drag.

Significant reduction in the wall friction is hence obviously technically possible using hydrophobic, flock-coated surfaces. Whether, however, this approach can be used under real conditions for ships depends on the stability of such a surface, the effect of this surface on the fouling behaviour, and whether the flock-coating is possible under shipyard conditions.

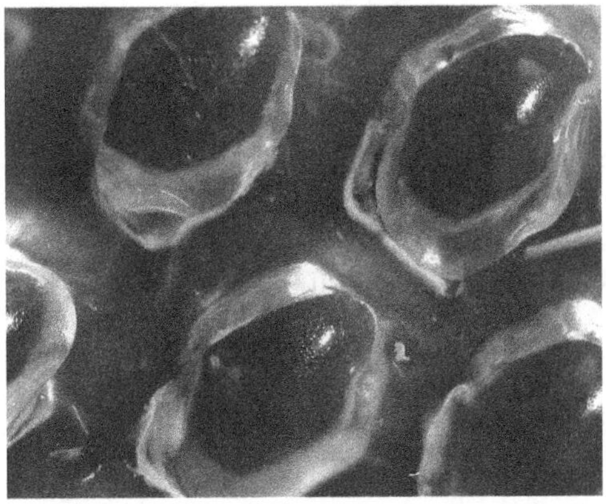

Figure 3.53: Immersed hydrophobic coating with honeycomb structure source: Fraunhofer IFAM

Hydrophobic surfaces with orderly structures

Surfaces with orderly, hydrophobic structures are an alternative to flock-coated surfaces. These are designed in such a way that the air is bound to the surface underwater, and can therefore induce its lubricating effect to reduce the drag. Honeycomb structures in a coating made hydrophobic using fluorinated binders have proved to be very effective for this. Figure 3.53 depicts such a surface with a honeycomb diameter of ca. 1 mm which has been immersed in water. It can clearly be seen that the air is held firmly in the honeycomb structure.

Such structured coating surfaces can, for example, be produced using the method described in Chapter 2.1.4.

The following points apply for both aforementioned approaches for air-trapping surfaces:

- Under high flow velocities, at least some of the air is removed from the surface.
- After a few days in water, the air that was initially trapped at the surface dissolves completely in the water. In order to use the effect over a longer period, the surface must therefore be replenished with air at regular intervals.

Air-trapping surfaces are hence generally able to reduce the drag of surfaces in water. Having said that, the practical application of this technology for commercial shipping, for example, is however still a long time away. The application of this technology in other areas, for example for sport equipment, is however generally promising.

3.4.5 Compliant coatings (Kramer-type coatings)

The skins of dolphins have unique flow properties: Although dolphins swim very fast (up to 10 knots), laminar flow is predominant around their bodies. This not only means very low drag, but also significantly reduced noise. From early on it was suspected that special mechanical properties of the skins of dolphins played a key role here.

In the late 1950s a series of experiments was undertaken which concluded that soft coatings inspired by the properties of dolphin skins were able to significantly reduce the drag of surfaces in water [16, 17]. These coatings have since that time been known as Kramer-type coatings. The effect of such coatings is based on the fact that the soft surface dampens the production and development of Tollmien-Schlichting waves (see above) via energy absorption (damping) and means that the switchover from laminar flow to turbulent flow occurs later. This leads to drag reduction because a larger portion of the surface is subject to laminar flow which is the more favourable condition.

This explanation appeared plausible to start with, but this effect could not be reproduced in later studies and so doubts about the results arose. Subsequent studies [18, 19] have however shown that Kramer-type coatings are theoretically able to reduce drag, but the mechanism is considerably more complex than outlined above.

Detailed experimental and theoretical work in the 1980s [20] suggested that these types of soft coatings have a high potential for reducing drag on ships and boats. No percentage drag reduction values can however be derived from these studies. In the 1990s more comprehensive investigations were undertaken. These were not restricted to the actual skin of the dolphin (ca. 2 mm thick), but also covered the fat layer below the skin (blubber) which is about 3 cm thick [21, 22]. This layer appears to have a key influence. It was shown that the viscoelastic properties of the blubber are such that they are perfectly adapted to the frequencies and amplitudes of the excitations caused by the turbulent boundary layer of the water. Synthetic materials, having comparable viscoelastic properties to the blubber, were also studied in this work. These materials included elastomeric polyurethane, a PVC (polyvinyl chloride) plastisol, and a 10 % PVC-dimethyl thianthrene gel. The mechanical properties of the latter are almost identical to those of natural blubber. It was examined how effectively the kinetic energy from the turbulent boundary layer is absorbed by these materials. In the case of the PVC-dimethyl thianthrene gel, 79 % of the energy was absorbed (for comparison: 1 % for polyurethane rubber and 1.5 % for PVC plastisol) [21, 22]. This is a strong indication that the concept of energy absorption from the boundary layer is actually the key effect of the dolphin skin resulting in reduced drag and noise.

Even following these studies, there is still a lack of proof that technical application is viable. A definite fact, however, from both the theoretical and experimental studies is that soft coatings delay the switchover point from laminar to turbulent flow and hence can contribute to reducing drag [23]. For this reason, this approach appears worthy of further investigation for technical applications.

3.4.6 Literature

[1] *Fukuda, K.; Tokunaga, J.; Nobunage, T.; Nakatani, T.; Iwasaki, T.; Kunitake, Y.*, Frictional drag reduction with air lubricant over a super-water-repellent surface, J. Mari. Sci. Technol., 5, 2000, 123-130

[2] *Bechert, D.W.; Bruse, M.; Hage, W.; Meyer, R.*, Fluid mechanics of biological surfaces and their technological application, Naturwissenschaften, 87 (4), 2000, 157-171

[3] *Bechert, D.W.; Hoppe, G.; Reif, W.-E.*, On the drag reduction of the shark skin, AIAA Shear Flow Control Conference; Boulder, CO, 1985

[4] *Coustols, E.; Savill, A.M.*, Résumé of important results presented art he third turbulent drag reduction working party, Applied Scientific Research, 46 (3), 1989, 183-196

[5] *Jang, P.S.; Benney, D.J.; Gran, R.L.*, On the origin of streamwise vortices in a turbulent boundary layer, Journal of Fluid Mechanics, 169, 1986, 109-123

[6] *Hage, W.*, Zur Widerstandsverminderung von dreidimensionalen Riblet-Strukturen und anderen Oberflächen Dissertation Berlin 2004

[7] *Bechert, D.W., Bruse, M.; Hage, W.*, Experiments with three-dimensional riblets as an idealized model of shark skin, Experiments in Fluids, 28, 2000, 403-412

[8] http://www.nasa.gov/centers/langley/news/factsheets/Riblets.html

[9] Jane's: All the World's Aircraft 1997-1998, Jane's Information Group, Coulsdon, Surrey, England (1997)

[10] *Stenzel, V.; Hage, W.*, Strömungsgünstige Beschichtung für die Senkung des Treibstoffverbrauchs, Tagungsband zum Ulmer Gespräch 2009, 72-79

[11] *Ou, J.; Rothstein, P.;* Direct velocity measurements of the flow past drag-reducing ultrahydrophobic surfaces, Physics of Fluids, 17, 2005, 103606

[12] *Ou, J.; Perot, B.; Rothstein, P.*, Laminar drag reduction in microchannels using ultrahydrophobic surfaces, Phys. Fluids, 16 (12) 2004, 4635-4643

[13] *Min, T.; Kim, J.*, Effects of hydrophobic surface on friction drag, Phys. Fluids, 16 (7) 2004, 55-58

[14] *Fukada, K.; Kasagi, N.*, A theoretical prediction of friction drag reduction in turbulent flow by superhydrophobic surfaces, Physics of Fluids, 18, 2006, 051703

[15] www.mikroblasen.de

[16] *Kramer, M.O.*, Boundary layer stabilization by distributed damping, J. Aeronaut. Sci., 24(6) 1957, 459-460

[17] *Kramer, M.O.*, The dolphin`s secret, Nav, Eng.J., 73 (1), 1961, 103-107

[18] *Carpenter, P.W.; Garrad, A.D.*, The hydrodynamic stability of flow over Kramer-type compliant surface. Part 2. Flow-induced surface instabilities, J. Fluid Mech. 170, 1986, 199-232

[19] *Benjamin, T.B., ed. H. Görtler*, Fluid flow with flexible boundaries, Proc. Int. Congr. Appl. Maths, 11th, 1964, Berlin, Springer-Verlag, p. 109-128

[20] *Riley, J.J.; Gad-el-Hak, M.; Metcalfe, R.W.*, Compliant Coatings, Ann. Rev. Fluid Mech., 20, 1988, 393-420

[21] *Fitzgerald, E.R.; Fitzgerald, J.W.*, Blubber and compliant coatings for drag reduction in water I. Viscoelastic properties of blubber and compliant coatings materials, Materials Science and Engineering:C2, 1995, 209-214

[22] *Fitzgerald, J.W.; Fitzgerald, E.R.; Carey, W.M.; Von Winkle, W.A.*, Blubber and compliant coatings for drag reduction in water II. Matched shear impedance for compliant layer drag reduction, Materials Science and Engineering: C2, 1995, 215-220

[23] *Carpenter, P.W.; Thomas, P.*, Flow over compliant rotating disks, J. Eng. Math., 57, 2007, 303-315
Van Dyke, M.: An album of fluid motion, The Parabolic Press, Stanford, California, USA, S. 93 (1982)

3.5 Anti-fouling coatings

Figure 3.54: Left: image of an underwater test stand for field experiments (source: G. Schmidt, Centre for Tropical Marine Ecology, Bremen, Germany); right: example of marine macro-fouling on a paint surface

source: Fraunhofer IFAM

Anti-fouling coatings suppress biological growth on surfaces and so prevent adverse effects on the relevant object or process. Microorganisms are generally responsible for biological growth. The consequences of microbial growth are varied, and include medical biofilm formation, mildew in living areas, microbial contamination of air conditioning equipment and systems through which water passes, and technical system faults caused by, for example, the blockage of filters and pipe systems. Besides microorganisms, macroorganisms such as barnacles and mussels are also responsible for biofouling. Figure 3.54 shows examples of biological growth in a marine environment.

Biofouling is costly, because the relevant surfaces have to be regularly cleaned and maintained. The hulls of ships are a good example of this. Here the biological growth also

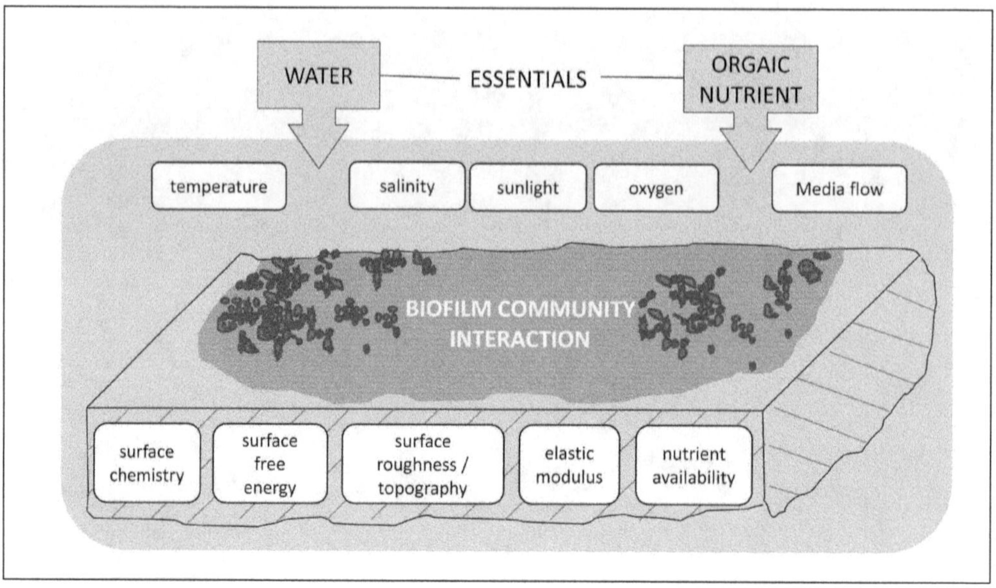

Figure 3.55: Factors affecting the formation and composition of the biofilm community: Besides essential factors such as water and organic nutrients, the environmental conditions and substrate properties are decisive

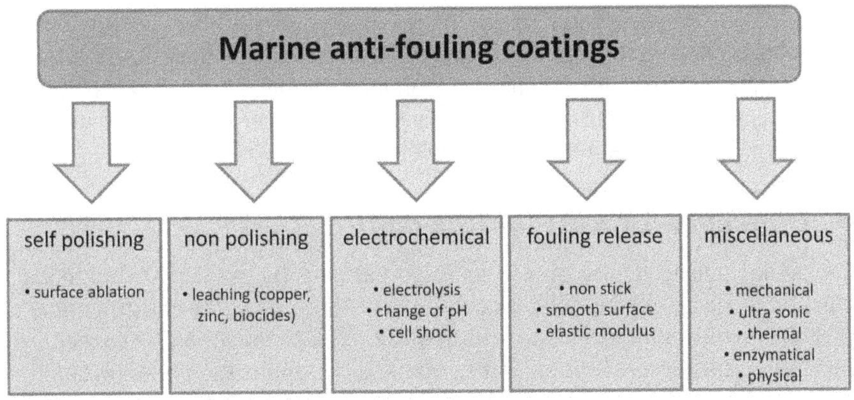

Figure 3.56: Concepts of marine anti-fouling coatings *source: Fraunhofer IFAM*

causes significantly higher fuel consumption due to the greater drag. A further possible consequence of biofouling on surfaces is microbially induced corrosion. In general, biofouling is accompanied by material damage, production downtime, quality loss, increased costs and work time, and last but not least potential harm to human health. This emphasises the commercial need for effective methods for combating biological growth (and these methods include anti-fouling coatings).

Virtually all surfaces are covered with microorganisms. The resulting biofilm forms an extremely heterogeneous and dynamic biofilm structure, consisting of living and dead cells as well as a layer of extracellular polymeric substances (EPS). This is formed by the microorganisms themselves and consists of, amongst other things, polysaccharides, proteins, and nucleic acids [1]. The list of organisms that can be present is long; in addition to microorganisms (viruses, bacteria, fungi, algae, diatoms, and protozoa), macroorganisms can also colonise surfaces (for example barnacles, mussels, annelida, and seaweed in a marine environment). The biofilm composition depends on many factors including the substrate, environmental conditions, and competitive interactions within the biofilm community (Figure 3.55). The large number of factors and the associated diversity of the biofilm mean that there is no single universal solution to combat biofouling on surfaces. Indeed, customised solutions are required.

A variety of methods are currently being used to combat biofouling. Solutions of biocidal agents are often used to kill microorganisms in industrial plants. There must then be a cleaning step (mechanical or chemical) because the dead cells can promote renewed biofouling. A further measure used for closed systems is to limit the supply of nutrients which, along with water, are essential for the formation of a biofilm. This is achieved, for example, using special filtering systems. A further method for fighting biofouling is to employ anti-fouling coatings, and these will be described in this section. According to DIN EN ISO 4618, anti-fouling coatings are "...coating material applied either to the underwater sections of a ship's hull or to other underwater structures to discourage biological growth thereon" [2]. This standard covers coatings for combating industrial and marine biofouling. Figure 3.56 summarises concepts of anti-fouling coatings for marine applications. Also included in this section will be coatings used in the hygiene and medical areas which are becoming ever more omnipresent (e.g. in fridges and functional clothing).

The following sections describe conventional and new concepts for anti-fouling coatings. The mechanisms of the coatings and the effects on the biofilm are explained. Brief biological background information is given in order to provide a general understanding of the strategies that can be employed for coatings. Outside the scope of this book, however, is detailed discussion of the often complex mechanisms of, for example, biocidal agents and the various interactions between organisms and surfaces. Readers seeking further information on these topics can, for example, refer to the book by Flemming et al. (2008) [3].

3.5.1 Concepts for anti-fouling coatings

Biofouling is the undesired growth of biological organisms (both microscale and macroscale organisms) on surfaces which deleteriously affects the properties and functions of those surfaces. It is only necessary to take measures when the negative effects exceed a tolerable level. Such a measure to prevent biofouling may be the application of an anti-fouling coating. Besides the term anti-fouling coating, a number of other terms for biologically active coatings are used in the literature (see Table 3.4).

The strategies outlined in Table 3.4 are employed in modern anti-fouling coatings. However, a combination of strategies is often required to reach optimum effectiveness. Decisive is that the selection of strategies is suitable for the relevant place of application and the relevant fouling-organisms. For example, a foul-release coating only offers a small chance of success in a static system because this requires the help of a flow (e.g. water flow) in order to be able to remove the organisms.

In principle, a distinction can be made between two concepts for anti-fouling coatings: biocide-containing coatings and foul-release coatings. Most of the biocide-containing are based on the leaching process. This involves active toxic substances being leached out of the coating surface and ending up on the surface. Perhaps the best known (and also most

Table 3.4: Summary of terms used in the literature to describe anti-fouling coatings (these terms are not mutually exclusive, indeed there is often overlap, for example a contact killing coating is often a biocidal coating).

Terms for anti-fouling coatings		
Target organism	Coating strategy	
Anti-microbial	Biocidal/Biostatic	Coatings that kill organisms/or inhibit the growth or proliferation of organisms
Anti-bacterial	Contact killing	Killing of organisms via contact with the coating (includes action of an anti-fouling agent)
Anti-fungal	Contact leaching	Dissolution of anti-fouling agent from paint matrix by surrounding media und release on the surface
	Self-polishing	Degradation of outermost polymeric layer due to chemical processes and subsequent exposure of “fresh” paint layer
	Self-eroding	Erosion of outermost polymeric layer due to mechanical stress and subsequent exposure of “fresh” paint layer
	Controlled depletion	Degradation of polymeric layer due to physical dissolution of paint matrix and its bioactive compounds
	Foul-release	Limited adhesion of fouling organisms due to unfavorable surface properties
	Repelling	Contact of the surface with water/contaminant/fouling organism without attachment to the coating

notorious) examples are tributyl tin self-polishing copolymer paints (TBT-SPC) that are used for ships. These coatings are not only toxic to target organisms on the surface but also to other aquatic organisms. Very low concentrations (ng/l level) of the TBT moiety cause abnormalities in aquatic organisms and lead to accumulation in mammals [4]. These negative effects resulted in a ban on TBT-containing coatings for ships. Alternative products, some containing zinc and copper compounds, are used today but these are not as effective as TBT-SPC. This has led to the use of organic-based booster biocides ("Irgarol" 1051 and "Diuron"). The use of these coatings is however also controversial due to the emission of metal ions into the environment and the toxicity of the booster biocides. These products can therefore only be considered to be short-term solutions [5]. Chapter 3.5.2 discusses biocide-containing coatings in greater detail and covers currently used coatings and novel products.

A second coating concept for preventing biofouling concerns the so-called repelling or foul-release coatings. The surface properties of these coatings either prevent the settlement of organisms or so strongly reduce the adhesion force that organisms can be easily removed. Key factors for foul-release coatings are their

- surface free energy,
- elastic modulus, and
- surface roughness.

The composition of these coatings is very complex and this matter is discussed in Chapter 3.5.3. Common to all these coatings is that they cannot hinder growth on static surfaces, but allow easier removal via shear and tensile stress. Trials on ships' hulls have demonstrated that macrofauna and macroflora can be removed at the sailing speeds typical of ships [5].

Only very general statements can be made about the effectiveness of the coating concepts. The large number of factors, the specific nature of the technical application, and the relevant organisms form a complex system and tailored anti-fouling coatings are required for optimum effectiveness. This complexity is also reflected in the large number of conceptual coatings and wealth of scientific studies. A detailed overview of the results of studies on chemical and physical coating parameters is given, for example, by Whitehead et al. (2008)[6]. Further examples are also described in detail in the chapters which now follow.

3.5.2 Biocide-containing coatings

The introduction of biocides into coatings is a very practical solution, because after application of the coating no further anti-fouling measures are required over its lifetime (assuming the dead biomass is removed by wear). On introducing anti-fouling agents into coating systems, the choice of coating matrix is important and attention must be paid to the ambient conditions such as the temperature, pH, and salinity at the point of action [6]. In addition, the compatibility of the components must be assured in order to realise a coating with satisfactory properties. The anti-fouling agents must not be chemically deactivated and release from the matrix must also be guaranteed. If the active substances were only to slowly leach out of the matrix, no adequate anti-fouling effect would be achieved. Conversely, if the active components are released too quickly, the service life of the coating is very limited.

Two different strategies are used in traditional biocide-containing coatings. Self-polishing coating systems are widely used and involve continuous release of the coating matrix in addition to the biocidal agent (Figure 3.57). The most successful (albeit now banned) representative of such systems are TBT-containing coatings for ships. These are based on an acrylic polymer (usually methyl methacrylate) with TBT groups bonded onto the polymer backbone by an ester linkage. The TBT moiety can be released by hydrolysis under slightly alkaline conditions (marine environment). The biocide is first released from the

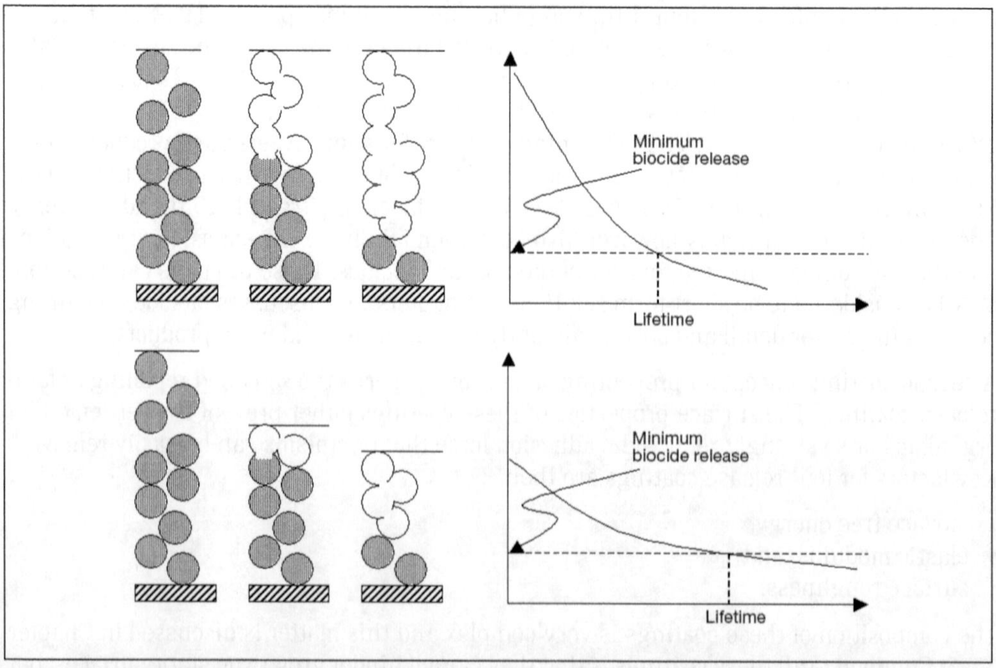

Figure 3.57: Mechanism and biocide release rates of insoluble and soluble matrix paints. "Minimum biocide release" indicates the limit for efficient protection against fouling[4].

Source: Elsevier Ltd., Oxford, UK; S. Kiil, Technical University of Denmark

uppermost region. In the resulting pores there is then a reaction between the coating matrix and the seawater. This leads to the degradation of the organic binder. As a result, fresh coating material is made available and the functionality is maintained over a longer period of time. Due to the TBT ban, self-polishing coatings based on copper and zinc have in the meantime come into use. These function by the same principle but are less effective and have a shorter service life [4].

In contrast, the second type of coating has an insoluble matrix. In these systems only the biocidal agent is released. The effective service life of these coatings is considerably shorter than for self-polishing coating systems (Figure 3.57). The advantage of these systems, however, is their higher mechanical stability. Typical representatives include insoluble vinyl, epoxy, acrylic, and chlorinated rubber polymers [4].

3.5.2.1 Biocides and mode of action

Biocides (or more precisely biocidal products) are defined in Directive 98/8/EC of the European Parliament as: "Active substances and preparations containing one or more active substances, put up in the form in which they are supplied to the user, intended to destroy, deter, render harmless, prevent the action of, or otherwise exert a controlling effect on any harmful organism by chemical or biological means" [7]. Biocides must not be harmful for people or the environment, and for many substances this is contradictory to their actual use (the killing of organisms). For this reason, and as already mentioned in the previous chapters with regard to TBT, many biocides have already been banned or are suspected of having harmful effects on people and the environment.

In order to be able to understand the mechanism of anti-microbial biocides, some knowledge of the cellular structure of microorganisms is required. In general, all cells are separated from their environment by cell membranes. Such a membrane consists of a double layer

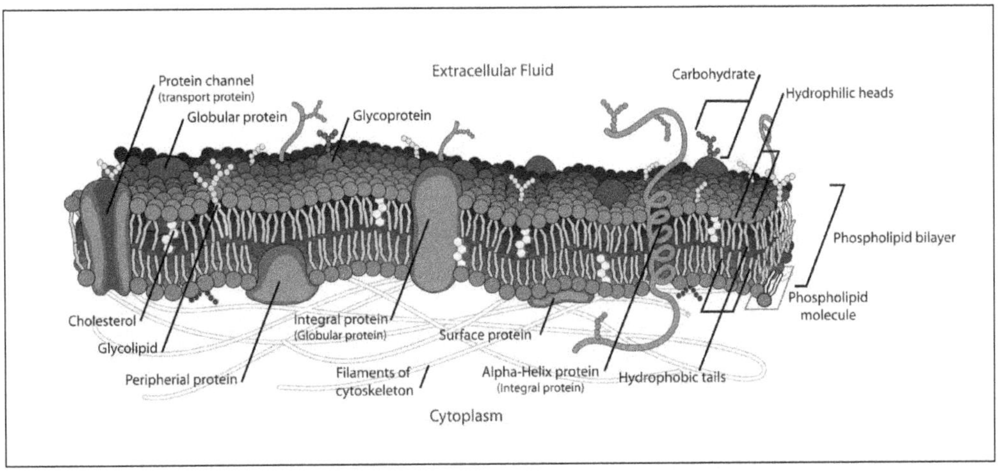

Figure 3.58: Scheme of a biomembrane *source: www.wikipedia.org, visited 10-01-2010*

of phospholipids, with both a hydrophobic and a hydrophilic part. A variety of protein molecules that act as channels and pumps for different molecules are embedded in this membrane (see Figure 3.58). Besides this separating function, cell membranes regulate transport mechanisms and maintain the electric potential of the cells. Some biocides attack the cell membrane directly whilst others are able to overcome these barriers and end up in the cell interior. As cell membranes are semi-permeable barriers, some molecules (in particular small molecules) can freely diffuse through the membrane whilst others can only get into the cell interior via specific transport molecules. The possible points of attack of biocides are therefore diverse, and this is reflected in the wealth of agents that are used. The effect is determined by the chemical structure of the biocide. At a cellular level, biocides cause amongst other things destruction of cell membranes, loss of ion gradients, damage to enzymes, and/or disruption of metabolic reactions [8].

Biocide mechanism of TBT

Having now briefly explained the biological mechanisms and structures, one can now turn to the effects of the biocides that are used in coatings. TBT (Figure 3.59) is a broad spectrum biocide and affects not only microorganisms but also higher organisms and mammals. Its mechanism is extremely complex and not well understood. For example, studies of Chicano et al. (2001) showed that organotin compounds react with phospholipids in the cell membranes and affect their function [9]. However, the emission of TBT into the environment caused undesired effects in non-target aquatic organisms. For numerous species of animals this led to the development of imposex and inability to reproduce, with some species of animals today threatened with extinction. In addition, bioaccumulation in higher animal species right up to mammals was demonstrated [4].

Copper and zinc based biocides

In general, a distinction can be made between two groups of biocides: non-metallic compounds and metal-based compounds. The latter include copper and zinc based biocides that are nowadays often used in coating systems. Both copper and zinc are essential elements that are required by organisms for maintenance of their life functions. If the concentrations of these metals, however, exceed a critical value, the metals have a toxic effect. Zinc compounds are, for example, used in coatings as broad-spectrum anti-microbial agents (e.g. zinc pyrithione) (Figure 3.59). These agents interact with ion-pumps of the bio-membranes and disrupt the transport mechanisms out of and into the cells [10]. With regards to their

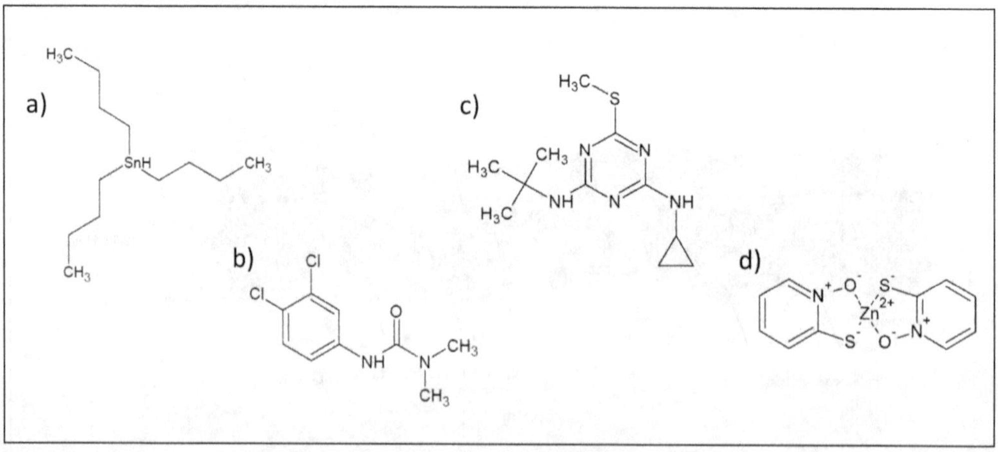

Figure 3.59: Examples of biocides: a) TBT, b) Diuron, c) "Irgarol" 1051, and d) zinc pyrithione

environmental compatibility, they have the advantage that there is rapid degradation in sunlight to less toxic products. They also have excellent biodegradability and there is little accumulation in sediments of surface waters. However, negative effects on the environment due to the accumulation of zinc cannot be excluded here [4].

Copper is common in the environment. If, however, the concentration of copper ions exceeds a critical value, these ions have a toxic effect on many aquatic organisms. Here, the sensitivity of the organisms varies widely and work of Yebra et al. (2004) indicates the following sensitivity sequence:

microorganisms > invertebrates > fish > bivalves > macrophytes [4].

The copper ions are able to pass through the cell membrane and at high concentrations destroy the functioning of enzymes in the cells [11]. Copper-containing pigments are used today in many anti-fouling coatings for marine applications in conjunction with other so-called booster biocides.

Booster biocides

Booster biocides are mostly non-metallic compounds and they intensify the toxic effect of the anti-fouling coating. For example, "Irgarol" 1051 (2-methyl-thio-4-tert-butylamino-6-cyclopropylamino-s-triazine) and "Diuron" (3-(3,4-di-chlorophenyl)-1,1-dimethylurea), for both see Figure 3.59, have been used and are being used as replacements for TBT. As, however, these biocides have also been demonstrated to have adverse environmental effects, their use has in the meantime also been banned or restricted [12].

These few examples already illustrate the issues concerning the use of both conventional and also newly developed biocidal coatings. The uncertainty regarding environmental effects means that long-term usage of these types of coatings is unlikely. In this connection, Evans et al. (2000) summarised the following uncertain factors [13]:

- environmental profiles of booster biocides;
- acute and chronic toxicity;
- validation of analytical methods for biocides, and monitoring the fate and toxicity in the environment;
- synergetic interactions between pollutants;
- accumulation in the environment;
- evaluation of the performance of alternative anti-fouling agents.

Despite the mentioned drawbacks, the huge commercial need for effective anti-fouling coating systems is resulting in a constant stream of new developments and scientific studies and these are described in the following chapter.

Anti-microbial coatings for medical applications

The examples mentioned up until now have concerned marine and industrial biofouling. Other important application areas include medicine and hygiene. Here, however, the terms anti-microbial coating and anti-bacterial coating are used. Such coatings are used, for example, on medical devices, implants, catheter surfaces, wound dressings, and also on textiles, food packaging, and even tooth brushes. For medical applications, antibiotics are often used as the active agents. The problem of resistance development emerged here (although this is not a problem limited to antibiotics), meaning that alternative active agents were necessary. Silver is an effective alternative agent. Even low concentrations inhibit microorganisms, by reacting with DNA structures within the cells and so preventing cell division. Silver also reacts with structures of the cell membrane and blocks key enzyme functions, so that the metabolism comes to a standstill [14]. The anti-microbial effect of silver is long known. In the past much effort has been put in to make the effect of silver coatings more efficient and more durable. Coatings with silver nanoparticles have also been developed and more information about this is given in the next section.

3.5.2.2 Alternative approaches

The literature contains many studies aimed at identifying alternative biocidal agents that have no harmful effects on the environment. The goal here is to develop coatings with at least an equivalent anti-fouling effect, but without the adverse effects for the environment. In order to give an insight into the future potential of biocidal anti-fouling coatings, this section outlines a selection of novel approaches. Naturally not all approaches can be covered here, and the focus is on selected examples which from a current perspective appear of interest for future applications.

Biometic strategies

A huge amount of scientific work is being put into the development of anti-fouling coatings that contain biocides which mimic what happens in nature. Many marine organisms have developed strategies to protect themselves against fouling. Besides physical and mechanical strategies, these also include the release of biocidal compounds and even giving shelter to bacteria with anti-fouling properties. For example, seaweeds accommodate bacteria on their surface that release anti-fouling compounds and so prevent settlement of other organisms [15].

The following steps have to be taken in order to develop alternative anti-fouling coatings containing biocidal components [4]:

- identification of the anti-fouling agent;
- incorporation of the bioactive substance(s) into the coating matrix without loss of effectiveness of the anti-fouling agent;
- compliance with coating requirements regarding mechanical properties, stability, and release characteristics;
- commercial competitiveness (costs and efficiency).

In addition, extensive approval procedures associated with the use of new biocides must be undertaken (see, for example, Olsen et al. (2007) [16]). However, the huge commercial benefits of effective anti-fouling coatings mean that some research groups still want to take on these challenges.

The bio-mimetic approaches include the use of secondary metabolites and enzymes, and the introduction of living bacteria into coating systems. These approaches, along with examples, are described below. Chapter 3.6 gives further examples of bio-mimicry.

Secondary metabolites

Secondary metabolites are substances synthesised by organisms that are not required for assuring their survival, growth, or reproduction. Rather, these substances often have specialised functions, for example for defence against competitors. A number of classes of substances come into consideration as anti-fouling agents, for example terpenoids, alkaloids, sterols, and fatty acids [5]. Scientific studies on anti-fouling effects have been carried out using metabolites extracted from organisms and synthetic analogs. Burgess et al. (2003), for example, investigated anti-fouling effects of bacteria isolated from living surfaces of seaweed. Bacterial extracts were formulated into water-based paints. Some showed anti-fouling properties in laboratory tests, and it was very clear that their introduction into a paint system can have a negative effect on the activity [17].

With regards to secondary metabolites, studies on polymeric 3-alkylpyridinium salts (poly-APS) and furanones are worthy of mention (see Chapter 3.6 "Bio-mimetic surfaces" for more details).

Enzymes as biocidal component

The use of enzymes for anti-fouling has also been investigated in numerous studies. Enzymes are proteins which catalyse (namely accelerate) chemical reactions. They can either act directly as a biocidal component or indirectly by aiding the formation of the anti-foulant. In addition, Olsen et al. (2010)[18] describe an interesting approach involving an anti-fouling effect induced by the release of hydrogen peroxide from enzyme-containing coatings. The mechanism behind this is as follows: First of all the starches incorporated into the coating are converted to glucose by the action of glycoamylase (GA) (Equation 3.2). The glucose in turn is directly converted into hydrogen peroxide and gluconolactone with the aid of hexose oxidase (HOx) (Equation 3.3). Hydrogen peroxide is a strong oxidising agent and is often used as a disinfectant because of its anti-bacterial effect. After being formed, hydrogen peroxide degrades very rapidly into water and oxygen (Equation 3.4) [18].

Equation 3.2 $(C_6H_{10}O_5)_n(s) + H_2O(l) \rightarrow (GA) \rightarrow C_6H_{12}O_6(aq) + (C_6H_{10}O_5)_{n-1}(s)$

Equation 3.3 $C_6H_{12}O_6(aq) + O_2(aq) \rightarrow (HOx) \rightarrow C_6H_{10}O_6(aq) + H_2O_2(aq)$

Equation 3.4 $2H_2O_2(aq) \rightarrow 2H_2O(l) + O_2(aq)$

Enzymes are rapidly biodegraded, meaning they can act as environmentally-friendly alternatives to metal-based anti-fouling agents. The feasibility of using enzymes in anti-fouling coatings has been studied for a long time. However, the authors are not aware of any commercial products that have emerged. Developing such coatings is complex, because many factors (e.g. stabilisation of the enzymes and release mechanisms) have to be guaranteed. This approach is described in more detail in Chapter 3.6. Olsen et al. (2007) give a detailed list of patents on enzyme-based anti-fouling coatings [16].

Living microorganisms

The third bio-mimetic approach that will be briefly described here is the introduction of living microorganisms into the coating matrix. This approach has a number of advantages, for example the production of biocidal compounds by the microorganisms that are introduced and the competition with undesired "settlers" for nutrients [16]. There are various studies and patents on this bio-mimetic approach and these are discussed in Chapter 3.6.

Encapsulated biocidal agents

In addition to the bio-mimetic approaches, work has been undertaken on other biocides and alternative release mechanisms. The latter include the use of encapsulated biocidal agents. This technology has already been described in Chapter 2.2. Microencapsulation offers better control of the leaching process and hence a longer coating lifetime. In addition, active agents which, for example, are not compatible with the coating formulation can be shielded so that satisfactory coating properties can be achieved and biocide release can occur as needed. For example, Fay et al. (2008) investigated the encapsulation of chlorohexidine with the biodegradable polymer poly(L-lactide) (PLA). Chlorohexidine is an antiseptic used in dentistry. Fay's research group found a promising approach for developing biodegradable anti-fouling paints based on non-toxic molecules and bioactive surfaces [19].

Hart et al. (2009) patented various types of encapsulation processes for use in marine anti-fouling coatings and paints. This patent describes single, dual, and multiple encapsulation processes. The resulting microcapsules range from 5 to 40 microns in size. By selecting the wall material, the permeability of the microcapsules in the marine environment can be controlled. Systems that are described include the encapsulation of isothiazolone biocides in a reaction product of polyvinyl alcohol and a phenolic resin, in a reaction product of an amino formaldehyde resin, and in a reaction product of polyvinyl alcohol and an isocyanate [20].

Other alternative release mechanisms for biocides can be found in the literature, for example electrolytically generated biocides (see [21]). These are, however, far removed from those of the organic coatings dealt with in this book and are hence not included here. Other examples are radioactive coatings and piezoelectric coatings which cannot be used due to their danger or are not feasible for large structures such as ships [4].

Hybrid coatings

However, an obvious alternative to conventional biocide-containing coatings is the combination of biocides with the foul-release concept described in the next chapter. These hybrid coatings should unite the benefits of the two anti-fouling concepts: the killing of organisms and the removal of residual organisms by minimising the adhesion force. Majumdar et al. (2008) studied, for example, hybrid coatings with tethered quaternary ammonium salts in a crosslinked polysiloxane matrix. The studies showed that this combination allows optimisation of the biocidal effect and an extended service life [22]. The foul-release concept is discussed in detail in the next chapter.

Silver containing coatings

The examples mentioned up until now have concerned marine and industrial bio-fouling. Other important application areas include medicine and hygiene, and the scientific work which has been undertaken to improve anti-fouling coatings in these areas will now be described. As already mentioned, the terms anti-microbial coating and anti-bacterial coating are often used here. An overview of the coatings used in the medical area has been given by, for example, Monteiro et al. (2009) [23]. Besides the use of various antibiotics, the use of silver must also be mentioned. The anti-microbial properties of silver have long been known. Polymers that release silver have shown strong anti-microbial activity [24]. However, conventional coatings show a high release rate (initial burst) after application, and the high concentration of released active agent can cause undesirable toxic reactions. For example in wound bandages or implants this can lead to tissue irritation and allergic reactions. In addition, the reservoir of silver can quickly become exhausted, meaning there is no longer satisfactory protection against infection. To combat these negative effects, a variety of concepts have been put forward. For example, coating systems have been developed that

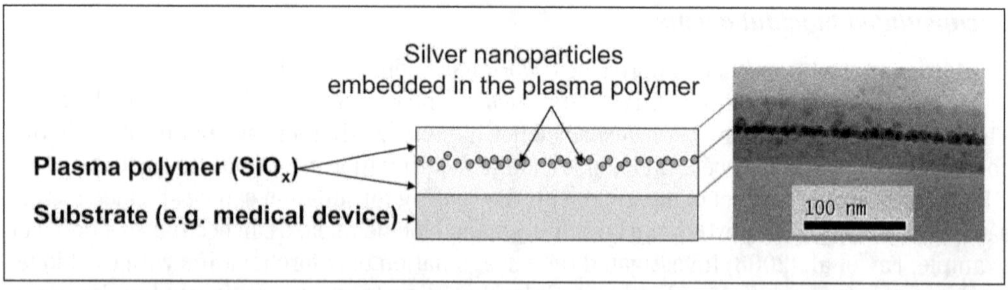

Figure 3.60: Scheme of layer composition of anti-bacterial plasma polymer coating with corresponding TEM-image *source: Fraunhofer IFAM and Bio-Gate AG*

contain a second, release-controlling layer. A patent [25] describes, for example, the use of a biocide-containing layer with a second, covering transport control layer (see Figure 3.60). The thickness and porosity of this second layer can be adjusted such that the biocide is released in an amount that has an anti-microbial effect and not a cytotoxic effect. The second layer can be applied using vacuum thin-layer methods, and in particular by sputtering and plasma-polymerisation [25].

Another concept for improving the effectiveness of silver coatings was investigated by the research group of Dowling. It was demonstrated that the addition of platinum improves the effectiveness, without having negative effects on the surrounding cell tissue. The mechanism behind this is enhanced oxidation of silver induced by the presence of the more noble platinum [26].

Use of nanoparticles

A further concept (the last that will be presented here) involves the use of nanoparticles. Nanoparticles, following the definition of the BSI (British Standards Institution) are "particles with one or more dimensions on the nanoscale", namely "...having one or more dimensions of the order of 100 nm or less" (PAS 71: 2005) [27]. These particles have a much greater surface-to-volume ratio than macroscale particles, meaning that a large surface is available for interaction with target organisms. This is the reason for their greater effectiveness compared to particles of conventional size, and means that the overall concentration of biocide can be reduced. However, the mode of action of, for example, silver nanoparticles is not well understood. The mechanisms discussed in the literature range from largely similar mechanisms to microscale particles (for example see [23]) right through to damage due to radical formation [28]. The properties of nanoparticles have been outlined in Chapter 2.3. Coatings containing silver nanoparticles are widely used today, and even in everyday objects such as computer keyboards and socks (to reduce sweaty odours). This widespread use of nanoparticles is however the subject of debate, given that there is a lack of knowledge about the potential risks to health.

In summary, it can be stated that each type of biocide-release coating described here is subject to statutory regulations and must meet the ever stricter requirements to minimise harmful effects on people and the environment. These must always be taken into account when appraising new product developments.

3.5.3 Foul-release coatings

Foul-release coatings facilitate the removal of residual organisms, and so prevent biofouling, by controlling the physical, chemical, and physico-chemical properties of surfaces. The concept involves reducing the adhesion force between settling organisms and the surface to such an extent that detachment of the organisms occurs due to their own weight or due to flow forces. Studies with foul-release coatings on the hulls of ships have demonstrated, for example, that macroorganisms can generally be removed at typical sailing speeds for ships (the reported minimum speed ranges from 7 to 22 knots). However, such coatings cannot currently remove microorganism-dominated slime films; indeed, even speeds of up to 30 knots are inadequate for this [4].

The removal of organisms using foul-release coatings is considerably enhanced with a flow (above all a water flow) and in many cases is only possible with a flow. The removal of fouling organisms merely due to their own weight, and effective prevention of settlement, has only been reported by a few authors in connection with specific macroorganisms such as barnacles [29, 30]. A further special case concerns surfaces that are only wetted with water occasionally and shorten this period due to their hydrophobic properties, and hence minimise fouling [31]. These systems are discussed in greater detail in the Chapter 3.5.3.4.

Foul-release coatings were developed in parallel with the TBT-containing self-polishing coatings, but did not make a breakthrough due to their poor effectiveness. R&D work in this area only gained impetus again after the ban and restrictions on biocide-containing coatings. The key advantage of foul-release coatings is their freedom from biocides, meaning that there is no danger to people and the environment from toxic substances.

A large number of factors influence the adhesion force between a substrate and fouling-organisms, namely the point of action of the foul-release coating. The following factors play a role here [32, 33]:

- chemical and physico-chemical properties of both the substrate surface and the organisms: surface energy, surface charge, hydrophobicity, surface chemistry;
- physical properties of the surface: roughness and topography;
- biological properties: species, strain, physiological state of the organisms, specific adhesion mechanisms such as fimbriae, pili, and exopolysaccharides;
- ambient conditions: for example ionic strength, pH, and nutrient availability.

The complexity of the surface properties and variety of bonding mechanisms of fouling-organisms mean that no general theory can be put forward for bioadhesion. As such there is not just one type of foul-release coating but rather surface coatings are developed for specific applications to provide protection against specific groups of organisms.

When considering the substrate properties it must also be remembered that although these properties are important for the initial cell attachment and adhesion, they have little influence after biofilm formation on the further biofilm development [6]. When a pure surface is immersed in water a conditioning film forms within a few minutes, consisting of organic molecules and inorganic material. This is the first change to the substrate surface. The conditioning film in turn forms the foundation for the microbial community that subsequently develops. This foundation changes constantly and can be settled upon by other organisms such as fouling macroorganisms [32]. The surface properties thus change constantly and this must always be taken into account when evaluating foul-release coatings.

The most promising foul-release coatings include coatings based on fluoropolymers, silicones, and polysiloxanes substituted with fluorine. Fluoropolymers, due to their very low surface free energy, have excellent non-stick properties. The disadvantage of these materials is the stiffness that is induced by the fluorine, meaning that the removal of attached organisms can be made more difficult. Silicones, due to their low elastic modulus, have a significant advantage over fluoropolymers here. In addition, silicone coatings are generally smoother, and this is also beneficial for the anti-fouling performance [4]. Silicone-based foul-release coatings for marine applications are currently available from a number of coating

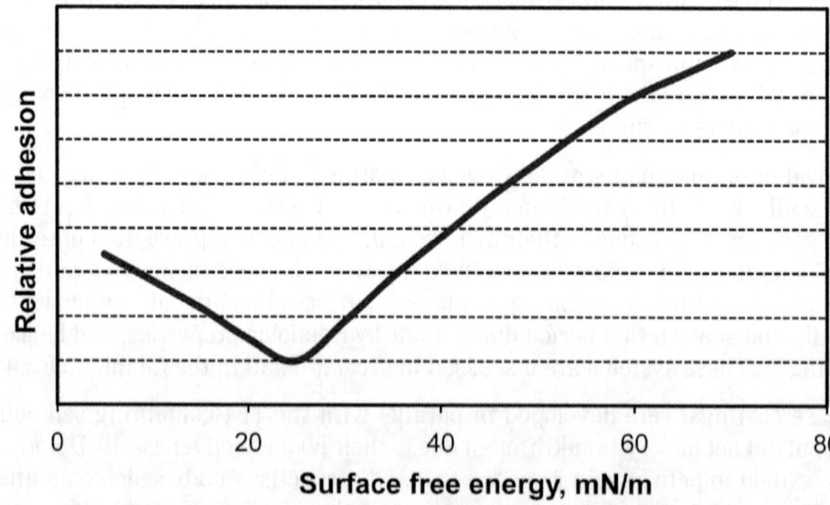

Figure 3.61: Baier curve [35] *source: C. Anderson, International Coatings*

manufacturers including International Paint Ltd. (UK), Hempel A/S (Denmark), and Sigma-Kalon Marine & Protective Coatings BV (Netherlands).

Improved foul-release performance was achieved by combining the elastic and smooth properties of silicone coatings with the very low surface energy of fluoropolymers. These systems contain polysiloxanes substituted with fluorine. A further possibility for improving the foul-release properties is the introduction of fluid additives (silicone oils) into the coating matrix which induces slippage at the surface [4].

The following chapters discuss the surface properties and their effect on the fouling in greater detail, based on research studies. It has already been mentioned that strict delineation of the surface parameters is often not possible because, for example, the surface chemistry and the surface roughness have an effect on the surface energy. This is taken into account in the following sections.

3.5.3.1 Surface free energy

The overriding current opinion is that the surface energy (= surface free energy) is the dominant substrate property affecting biological growth and the adhesion force of organisms. The surface energy here is a measure of the energy of free surface groups, molecules, and atoms on the substrate surface that can interact with approaching groups, molecules, and atoms. These interactions include van-der-Waals forces, polar and electrostatic interactions, and hydrogen bonding. The lower the surface energy, the weaker the adhesion, because the number of possible interactions is low. This hence means that low energy surfaces are advantageous for foul-release coatings because, due to the weak bonds (or weak interactions) at the substrate-liquid interface, these surfaces are less intensely colonised by organisms and the growth can be more easily removed [32].

The surface energy of a solid is indirectly determined via contact angle measurements of different liquids of known surface tension and calculated using Young's equation (for further information see, for example, [31] and see also Chapter 2.1.2.1). In addition to the surface energy, surfaces are often characterised by their hydrophobicity (water repelling, low wettability) or hydrophilicity (water attracting, high wettability) already discussed in Chapter 2.1 (e.g. Figure 2.4). In general, hydrophobic surfaces have low surface energies.

Low energy surfaces

A number of studies have shown that low energy surfaces have poor bioadhesion for most (but not all) fouling organisms. The lowest adhesion forces were determined for surface energies of between 20 and 30 mN/m [33-36]. These results confirm the profile of the so-called Baier curve (Figure 3.61) which indicates that the minimum adhesion (but not zero adhesion) between the surface and fouling organism occurs at 22 to 24 mN/m. Both lower and higher surface tension induce an increase in bioadhesion [34, 35]. The reasons for this observation were stated on the one hand as being due to the effect of the conditioning film and on the other hand due to the overlapping effects of elastic modulus, roughness, and coating thickness. Furthermore, the surface energy of the organisms themselves and the surface tension of the suspending liquid also have an influence [31, 35, 36].

The surface energy of a coating depends on several factors. The surface chemistry has a significant effect, because the selection of surface groups is decisive for subsequent interactions with approaching molecules and organisms. To create surfaces having as few as possible interactions, moieties with primarily methylated and fluorinated carbon atoms are used. The surface energy decreases in the following sequence:

$-CH_2- > -CH_3 > -CF_2- > -CF_2H > -CF_3$ [31].

A further factor affecting the surface energy is the surface roughness. Whilst water contact angles of maximum ~130° can be achieved for smooth surfaces, values of up to 170° can be achieved by customisation of the surface roughness (superhydrophobic surfaces) [31], see Chapter 2.1.2. The effect of surface roughness and surface topography on the anti-fouling properties is discussed again in Chapter 3.5.3.3.

Silicone-based systems

Hydrophobic coatings are often silicone-based systems and fluorinated polymer coatings. The majority of studies have given prominence to silicone-based coatings as being the most promising. Tests on these systems showed that fouling growth rates were lower, the bioadhesion was lower, and the surfaces were easier to clean (for example see ref. [32]). It was also shown that silicone coatings had only weak adhesion to fouling macroorganisms such as barnacles, tubeworms, and macroalgae [37].

Holm et al. (2006) studied the bioadhesion of various macroorganisms on 12 silicone fouling release surfaces. It emerged that none of the tested coatings had a minimum adhesive strength for all test organisms; coatings with good results for, for example, tubeworms were less effective for barnacles. They concluded that in order to develop globally effective foul-release coatings a wide spectrum of fouling organisms had to be considered [38]. In addition, there are organisms that prefer low energy surfaces and react with a larger cell-substrate contact area [32]. From these few examples, it is clear that given the diversity of fouling organisms it is a huge challenge to develop effective foul-release coatings.

The fact that silicone coatings came out on top in the studies is not only due to their surface energy, but rather due to the complex interplay with other factors such as the elastic modulus and roughness. These will be looked at in subsequent sections. It is important here to mention the slippage, the effect of which was investigated by Meyer et al (2006) on the adhesion of biofoulants. It was shown that the incorporation of non-functional silicone oils into the coating materials did not affect the surface energy, but did reduce the bioadhesion [39]. As such, friction and lubricity are also factors that influence bioadhesion.

3.5.3.2 *Elastic modulus*

In addition to the surface energy, the elastic modulus of a coating is also decisive for the bioadhesion of fouling organisms. In general, the elastic modulus describes the relationship between the stress and strain on deformation of a material and is a measure of its resistance to deformation. The greater the resistance of a material to deformation, the greater the elastic modulus. Regarding the anti-fouling properties of coatings, it was established that as low as possible elastic modulus (namely good, reversible deformation) has a positive effect on the bioadhesion [37]. This relates to the fact that elastomeric silicone coatings are deformed when a force acts and hence the bond between the coating and the marine adhesive is broken. This process is relatively slow compared to detachment processes for fluoropolymers, but requires a significantly lower energy [4]. A further effect observed in coatings with a low elastic modulus is reduced settlement by fouling macroorganisms (prevention of settlement). Natasha et al. (2002) also postulated that due to the low E-modulus there is mechanical deformation of the sensor membranes which are used by the larvae of a very wide number of fouling macroorganisms for testing the surfaces to be colonised. The result is that these surfaces are avoided [30].

Of the possible coating systems, siloxane elastomers are to the fore because these are very flexible compared to other polymers. This is due to their chemical structure with a silicon-oxygen-silicon backbone. In addition, the desired elastic modulus can be set via the degree of crosslinking. The balance between elasticity and stability must be found here, and this can differ for different applications.

A further parameter that should be briefly mentioned here is the thickness of the coating. It was found that a thickness of greater than 100 μm should be applied. This on the one hand prevents macroorganisms such as barnacles cutting through the coating and attaching to a firmer subsurface below [35]. On the other hand it improves the fracture mechanics at the interface between the coating and fouling organism [4].

3.5.3.3 *Surface roughness/topography*

The roughness of a surface can also partially dictate the fouling properties of a coating. For example, it affects the spreading of liquid cements secreted by organisms. If these adhesive cements cannot fill all the small crevices due to their high viscosity, the adhesion force is minimised. Conversely, if the contact surface is enlarged due to suitable roughness, stronger adhesion forces occur [37].

In general, the roughness and surface topography are assumed to only have a small effect on the growth of microorganisms. This is due to the prevailing size ratios: the surface features are generally much larger than most bacteria cells, meaning that they can only play a secondary role [6, 37].

Regarding fouling macroorganisms such as barnacles and tubeworms, the roughness can be considered in a different way. Various studies have investigated the effect of roughness/surface topography on bioadhesion and prevention of colonisation by organisms. Andersson et al. (1999), for example, demonstrated anti-fouling effects on microtextured, low surface energy RTV silicones (PDMS). The homogenous surface structures consisted of a mesh structure (in the 50 to 100 μm region). The so structured coatings showed significantly reduced barnacle attachment in field tests compared to smooth PDMS surfaces. Besides reduced adhesion due to the poor wetting of the adhesive cement, other reasons for the observed anti-fouling effect were also discussed. These included the fact that the size of the adhesive antennular discs of the barnacles was similar to the size of the surface microstructure, meaning that the exploring barnacle larvae avoid these surfaces [29]. Similarly, a systematic study looked at the effect of differently sized microtexture on colonisation by fouling organisms. It was demonstrated here that macrofouling larvae, in particular, have a lower affinity for surfaces whose microtexture is slightly smaller in size than the width of the organism (attachment point theory). In contrast, this relationship only applied to a limited extent, or not at all, for fouling microorganisms and non-motile organisms [40].

These results indicate that customised surface structuring is a way of suppressing colonisation by certain fouling macroorganisms. However, a globally effective anti-fouling coating cannot be achieved via the creation of artificial, homogenous structures on surfaces. Surface topographies that mimic nature may open further opportunities here. Studies on the effectiveness of such bio-mimetic surfaces are discussed in more detail in Chapter 3.6.

The introduction of surface structuring is relative complex from a technical point of view (see Chapter 2.1) and is hence relatively costly. Whether these costs justify the benefits must be determined for each individual application. In addition, questions about the lifetime of the surface structures and the effect of abrasion and erosion still have to be answered. Surface structuring currently plays only a fairly minor role as an anti-fouling strategy.

3.5.3.4 Other approaches

A variety of new approaches have been put forward for improving the properties and effectiveness of foul-release coatings. Already mentioned is the use of superhydrophobic surfaces, in particular for applications where surfaces are only occasionally exposed to water. The rapid water removal minimises the chances of "contamination". In general, non-wettability can be achieved by (see [31]):

- reduction of the contact surface between the substrate and liquid (caused by roughness);
- incorporation of fluorine.

Various approaches for this can be found in the literature [31]. One of these is worthy of mention here: Li et al. (2001) manufactured surfaces containing vertically standing carbon nanotubes (CNTs). These demonstrated marked hydrophobicity, and this was further enhanced by fluorination of the CNTs (water contact angle >170°) [41]. A further way of creating superhydrophobic surfaces is to mimic surfaces that are found in nature, such as the lotus leaf structure. This has already been discussed in Chapter 2.1 and 3.1, and Chapter 3.6 will also touch on this.

A totally different approach to the aforementioned is the creation of superhydrophilic surfaces. Hydrophilic surfaces in general lead to enhanced cell adhesion and cell growth. The optimum for biofouling interactions is a water contact angle of ca. 60° [37]. Superhydrophilic surfaces, however, have considerably lower water contact angles. The affinity for water molecules here is so strong that foreign substances are kept away from the surface. The surfaces that were studied showed low protein adsorption, with non-charged hydrophilic polyethylene glycol (PEG) surfaces giving the best results. Due to the low protein adsorption, the formation of a conditioning film is suppressed, which is the most important precondition for subsequent biofouling processes [37].

A wide range of wet chemical methods and plasma treatment can be used for modifying surfaces with PEG groups. For example, PEG chains can be manufactured as brush-type surface coatings using mono-functional PEG acrylates and UV polymerisation (also see Chapter 2.5). These surfaces have a water contact angle < 10° coupled with extremely low protein adsorption. PEG segments are also used for developing surface responsive materials, here in combination with fluorine compounds and PDMS. This combination allows surfaces to be generated that have non-stick properties at the air-solid interface due to the fluorine and silicone components. They have anti-fouling properties on exposure to water

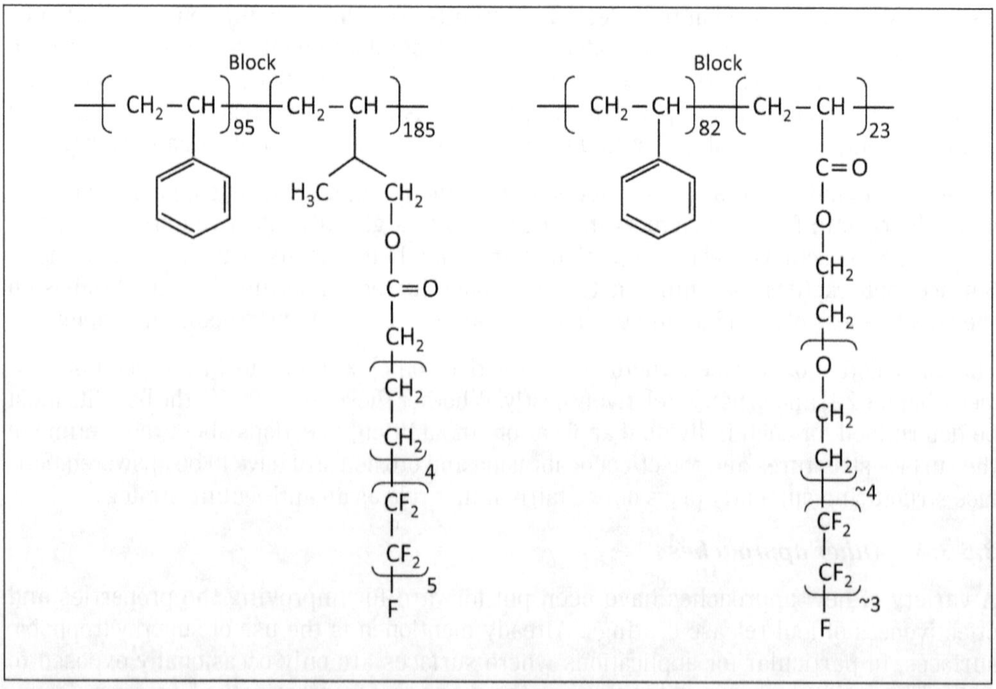

Figure 3.62: Example of block copolymers used for surface PEGylation [42]

due to the migration of the PEG segments to the water-solid interface. A detailed overview of these coatings is available [37, 42, 43].

Various research groups are developing superhydrophilic coatings. Studies on the anti-fouling effects have shown that combining the non-stick properties of non-polar fluoroalkyl groups with the anti-fouling characteristics of hydrophilic PEG groups resulted in reduced adhesion and reduced colonisation by various groups of organisms (results summarised by Krishnan et al. (2008) [42]). Even for the good results, however, it must be remembered that the long-term stability of these coatings in a biological environment is not a given. Oxidative degradation and chain cleavage mean that there is still uncertainty about the practical implementation of these anti-fouling coatings [42]. Further development work is necessary here.

The approaches that have been discussed here are all at the experimental stage and it is still difficult to estimate their promise as anti-fouling coatings for major applications.

3.5.4 Conclusions

Biofouling is a ubiquitous problem. Biofouling is costly to remove and suppress, and for these reasons much research work is focused on preventing the biofouling of surfaces. A very effective approach is to use biocide-containing coatings. These are able to kill a broad spectrum of fouling organisms and so protect technical equipment, artificial underwater structures (such as ships' hulls), and medical materials against fouling. There is much experience with these types of coatings and a wide range of products is available. However, this range of products will continue to shrink in the future because biocide-containing coatings not only kill the target organisms but also have negative effects on people and the environment. The result is ever more far-reaching bans on their use, which in turn is opening up opportunities for other anti-fouling strategies. These include foul-release coatings whose anti-fouling properties are due to customisation of surface properties such as the surface energy, elastic modulus, and topography. The systems which are currently available are, however, less effective and costly for end users. This is why much effort is being put in to improve the effectiveness of alternative anti-fouling coatings.

From the results described here, it can be stated in summary that in the medium term there is not likely to be a biocide-free solution of equivalent effectiveness for a wide range of applications. The diversity of organisms, environmental conditions, and required specifications leads one to the conclusion that customised solutions will become ever more prominent. For the development of these coatings, collaboration with a range of specialists including biologists, chemists, engineers, and paint specialists is necessary in order to find tailored solution for specific technical applications.

3.5.5 Literature

[1] *Flemming, H.C.*, Why Microorganisms Live in Biofilms and the Problem of Biofouling, In: *Flemming, H.C., Sriyutha Murthy, P., Venkatesan, R., Cooksey, K.E.* (Eds.), Marine and Industrial Biofouling, Springer-Verlag Berlin Heidelberg, 2008, pp.3-12

[2] DIN EN ISO 4618 Paints and varnishes - terms and definitions, 2007

[3] *Flemming, H.C., Sriyutha Murthy, P., Venkatesan, R., Cooksey, K.E.* (Eds.), Marine and Industrial Biofouling, Springer-Verlag Berlin Heidelberg, 2008

[4] *Yebra, D.M., Kiil, S., Dam-Johansen, K.*, Antifouling technology - past, present and future steps towards efficient and environmentally friendly antifouling coatings, Progress in Organic Coatings, 50, 2004, 75-104

[5] *Sriyutha Murthy, P., Venugopalan, V.P., Nair, K.V.K., Subramoniam, T., Larval* Settlement and Surfaces: Implications in development of Antifouling Strategies, In: *Flemming, H.C., Sriyutha Murthy, P., Venkatesan, R., Cooksey, K.E.* (Eds.), Marine and Industrial Biofouling, Springer-Verlag Berlin Heidelberg, 2008, pp.233-263

[6] *Whitehead, K.A., Verran, J.*, The Effect of Substratum Properties on the Survival of Attached Microorganisms on Inert Surfaces, In: *Flemming, H.C., Sriyutha Murthy, P., Venkatesan, R., Cooksey, K.E.* (Eds.), Marine and Industrial Biofouling, Springer-Verlag Berlin Heidelberg, 2008, pp.13-33

[7] EU, 1998, Directive 98/8/EC of the European Parliament and of the Council on biocidal products, In Article 2 (1a), Official Journal No. L123/1, 1998, ec.europa.eu/environment/biocides/pdf/dir_98_8_biocides.pdf

[8] *Nendza, M.*, StructureActivity Relationships in Environmental Sciences, Chapman & Hall, London, 1998

[9] *Chicano, J.J., Ortiz, A., Teruel, J.A., Aranda, F.J.*, Organotin compounds alter the physical organization of phosphatidylcholine membranes, Biochimica et Biophysics Acta 1510, 2001, 330-341

[10] *Chandler, C.J., Segel, I.H.*, Mechanism of the Antimicrobial Action of Pyrithione: Effects on Membrane Transport, ATP Levels, and Protein Synthesis, Antimicrobial Agents and Chemotherapy, Vol.14, No.1, 1978, 60-68

[11] *Mann, E., Ahlgren, N., Moffett, J.W., Chisholm, S.W.*, Copper Toxicity and Cyanobacteria Ecology in the Sargasso Sea, JSTOR Limnology and Oceanography, Vol.47, No.4, 2002, 976-988

[12] *Chambers, L.D., Stokes, K.R., Walsh, F.C., Wood, R.J.K.*, Modern approaches to marine antifouling coatings, Surface & Coatings Technology, 201, 2006, 3642-3652

[13] *Evans, S.M., Birchenough, A.C., Brancato, M.S.*, The TBT Ban: Out of the Frying Pan into the Fire?, Marine Pollution Bulletin, Vol.40, No.3, 2000, 204-211

[14] *Rogers, J., Dowsett, A.B., Keevil, C.W.*, A paint incorporating silver to control mixed biofilms containing *Legionella pneumophila*, Journal of Industrial Microbiology, 15, 1995, 377-383

[15] *Armstrong, E., Boyd, K.G., Burgess, J.G.*, Prevention of marine biofouling using natural compounds from marine organisms, Biotechnology Annual Review, Vol.6, 2000, 221-241

[16] *Olsen, S.M., Pedersen, L.T., Laursen, M.H., Kiil, S., Dam-Johansen, K.*, Enzyme-based antifouling coatings: a review, Biofouling, 23(5/6), 2007, 369-383

[17] *Burgess, J.G., Boyd, K., Armstrong, E., Jiang, Z., Yan, L., Berggren, M., May, U., Pisacane, T., Granmo, A., Adams, D.R.*, The Development of a Marine Natural Product-based Antifouling Paint, Biofouling, 19, 2003, 197-205

[18] *Olsen, S.M., Kristensen, J.B., Laursen, B.S., Pedersen, L.T., Dam-Johansen, K., Kiil, S.*, Antifouling effect of hydrogen peroxide release from enzymatic marine coatings: Exposure testing under equatorial and Mediterranean conditions, Progress in Organic Coatings, Vol.68, No.3, 2010, 248-257

[19] *Fay, F., Linossier, I., Legendre, G., Vallée-Rehel, K.*, Micro-Encapsulation and Anti-Fouling Coatings: Development of Poly(lactic acid) Microspheres Containing Bioactive Molecules, Macromol. Symp., 272, 2008, 45-51

[20] *Hart, R.L., Virgallito, D.R., Work, D.E.*, Microencapsulation of Biocides and Antifouling Agents, Patent US 2009/0186058 A1, 2009
[21] *Venkatesan, R., Sriyutha Murthy, P.*, Macrofouling Control in Power Plants, In: *Flemming, H.C., Sriyutha Murthy, P., Venkatesan, R., Cooksey, K.E.* (Eds.), Marine and Industrial Biofouling, Springer-Verlag Berlin Heidelberg, 2008, pp.13-33
[22] *Majumdar, P., Lee, E., Patel, N., Stafslien, S.J., Daniels, J., Chisholm, B.J.*, Development of environmentally friendly, antifouling coatings based on tethered quaternary ammonium salts in a crosslinked polydimethylsiloxane matrix, J. Coat. Technol. Res., 5 (4), 2008, 405-417
[23] *Monteiro, D.R., Gorup, L.F., Takamiya, A.S., Ruvollo-Filho, A.C., deCamargo, E.R., Barbosa, D.B.*, The growing importance of materials that prevent microbial adhesion: antimicrobial effect of medical devices containing silver, Int. Journal of Antimicrobial Agents, 34, 2009, 103-110
[24] *Hetrick, E.M., Schoenfisch, M.H.*, Reducing implant-related infections: active release strategies, Chem. Soc. Review, 35, 2006, 780-789
[25] *Wagener, M., Vissing, K.D., Salz, D., Steinrücke, P.*, Antimikrobielles Schichtmaterial, Patent EP 1 790 224 A1, 2007
[26] *Dowling, D.P., Betts, A.J., Pope, C., McConnell, M.L., Eloy, R., Arnaud, M.N.*, Anti-bacterial silver coatings exhibiting enhanced activity through the addition of platinum, Surface and Coatings Technology, 163-164, 2003, 637-640
[27] BSI - British Standards Institution: PAS 71:2005 Vocabulary - Nanoparticles, London, May 2005
[28] BfR - Bundesinstitut für Risikobewertung, BfR-Delphi-Studie zur Nanotechnologie - Expertenbefragung zum Einsatz von Nanomaterialien in Lebensmitteln und Verbraucherprodukten, Berlin, 2009
[29] *Andersson, M.; Berntsson, K.; Jonsson, P.; Gatenholm, P.;* Mictrotextured Surfaces: towards Macrofouling Resistant Coatings, Biofouling, 14 (2), 1999, 167-178
[30] *Natasha, L.G.; Banta, W.C.; Loeb, G.I.;* Aquatic Biofouling Larvae Respond to Differences in the Mechanical Properties of the Surface on which they Settle, Biofouling, 18 (4), 2002, 269-273
[31] *Genzer, J., Efimenko, K.*, Recent developments in superhydrophobic surfaces and their relevance to marine fouling: a review, Biofouling, 22 (5/6), 2006, 339-360
[32] *Callow, M.E., Fletcher, R.L.*, The Influence of Low Surface Energy Materials on Bioadhesion - a Review, International Biodeterioration & Biodegradation, 1995, 333-348
[33] *Pereni, C.I. Zhao, Q., Liu, Y. Abel, E.*, Surface free energy on bacterial retention, Colloids and Surfaces B: Biointerfaces, 48, 2006, 143-147
[34] *Baier, R.E.*, Surface behavior of biomaterials: The *theta surface* for biocompatibility, J Mater Sci: Mater Med, 17, 2006, 1057-1062
[35] *Anderson, C., Atlar, M., Callow, M., Candries, M., Milne, A., Townsin, R.L.*, The development of foul-release coatings for seagoing vessels, Journal of Marine Design and Operations, 2003, 11-23
[36] *Cooksey, K.E., Wigglesworth-Cooksey, B., Long, R.A.*, A Strategy to Pursue in Selecting a Natural Antifoulant: A Perspective, In: *Flemming, H.C., Sriyutha Murthy, P., Venkatesan, R., Cooksey, K.E.* (Eds.), Marine and Industrial Biofouling, Springer-Verlag Berlin Heidelberg, 2008, pp.165-177
[37] *Vladkova, T.*, Surface Modification Approach to Control Biofouling, In: *Flemming, H.C., Sriyutha Murthy, P., Venkatesan, R., Cooksey, K.E.* (Eds.), Marine and Industrial Biofouling, Springer-Verlag Berlin Heidelberg, 2008, pp.135-163
[38] *Holm, E.R., Kavanagh, C.J., Meyer, A.E., Wiebe, D., Nedved, B.T., Wendt, D., Smith, C.M., Hadfield, M.G., Swain, G., Darkangelo Wood, C., Truby, K., Stein, J., Montemarino, J.*, Interspecific variations in patterns of adhesion of marine fouling to silicone surfaces, Biofouling, 22:4, 2006, 233-243
[39] *Meyer, A., Baier, R., Wood, C.D., Stein, J., Truby, K., Holm, E., Montemarano, J., Kavanagh, C., Nedved, B., Smith, C., Swain, G., Wiebe, D.*, Contact angle anomalies indicate that surface-active eluates from silicone coatings inhibit the adhesive mechanisms of fouling organisms, Biofouling, 22 (6), 2006, 411-423
[40] *Scardino, A.J., Guenther, J., deNys, R.*, Attachment point theory revisited: the fouling response to a microtextured matrix, Biofouling, 24 (1), 2008, 45-53
[41] *Li, H., Wang, X., Song, Y., Liu, Y., Li, Q., Jiang, L., Zhu, D.*, Super-„Amphiphobic" Aligned Carbon Nanotube Films, Angew. Chem., 113, Nr.9, 2001, 1793-1796
[42] *Krishnan, S., Weinman, C.J., Ober, C.K.*, Advances in polymers for anti-biofouling surfaces, J. Mater. Chem., 18, 2008, 3405-3413
[43] *Russell, T.P.*, Surface responsive materials, Science 279, 2002, 964-967

3.6 Bio-mimetic surfaces

3.6.1 Introduction

Bionics is a field of science whose goal is to transfer natural processes and structures into technical solutions. The first step here is often identifying and investigating the strategies that are used in nature. The knowledge that is gained about the processes and structures must then taken from the natural context and converted into technical solutions. One of the best known examples of such a development is the transfer of the natural surface properties of lotus leaves to artificial surfaces in order to realise self-cleaning functionality (assisted by water). The identification and study of the lotus effect and the utilisation of these findings to produce artificial surfaces were and are an important step in the development of bionic solutions.

However, bionics covers much more that the mimicking of the surfaces of lotus leaves. Besides the functional surfaces which will be discussed here, the area of bionics also covers established fields such as construction (e.g. lightweight construction and flow optimisation) and developing fields such as prosthetics and nano-biomimetics (e.g. reproduction of photosynthesis) [1]. Regarding functional biomimetic surfaces, the following examples are discussed in detail in this book:

- Self-cleaning, superhydrophobic surfaces (lotus effect) - Chapter 3.1
- Drag-reducing surfaces - Chapter 3.4
- Optical surfaces (e.g. moth eye structures for anti-reflective properties)
- Bio-inspired adhesion (Gecko feet and DOPA-linked surfaces)
- Anti-fouling surfaces - Chapter 3.5, 3.6
- Anti-icing surfaces - Chapter 3.2, 3.6
- Self-healing surfaces -Chapter 3.3
- Surfaces with minimal solid friction - the sandfish

Some of these functionalities have been described in detail elsewhere in this book, and are merely referred to here. Other biomimetic approaches will be covered in the chapters which follow, although we do not claim to mention all biomimetic developments. A fundamental distinction can be made between structure-inspired and (bio)chemical-inspired surface functionalities and this is adopted in the chapters which follow.

Drag-reducing and self-cleaning surfaces, and also self-healing coatings, are considered to have a very promising future [1]. A few products are already in the marketplace which the manufacturers claim to be based on bionic principles. For example, a number of manufacturers produce low-drag swimming costumes that are based on the principle of shark skin and dolphin skin. In the area of self-cleaning surfaces, some exterior facade coatings and other coatings are being marketed as bionic. The other functionalities that are described here are of high interest, but development work is still at an early stage.

3.6.2 Structure-inspired functions

The structure of a surface largely determines its properties and is hence very important for manufacturing functional surfaces. Studying surface structures that occur in nature is very worthwhile because evolutionary processes have resulted in highly effective functional/multifunctional surfaces. Examples worthy of mention here are sharkskin with its very low-drag surface and lotus leaves with their dirt-repelling surfaces. Technical surfaces have been produced which are based on these natural surfaces and these have already been discussed in detail in previous chapters of this book: See Chapter 3.1 (dirt-repelling surfaces)

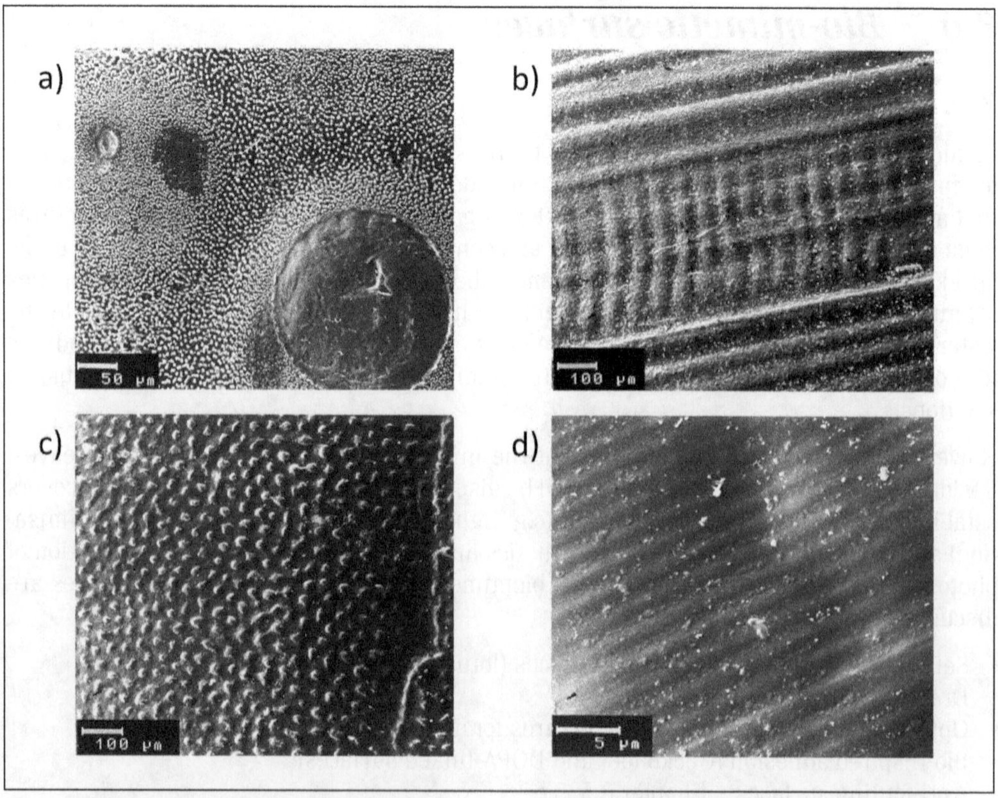

Figure 3.63: Surfaces of different marine organisms whose anti-fouling properties were studied (images kindly provided by V. Bers[4]) a) Microtopography of the edible crab - evenly distributed circular elevations ~200 µm in diameter; spicule-like structures between the elevations of length 2 to 2.5 µm; b) Surface structures of the egg case of the small-spotted catshark - longitudinal ridges with irregular spacing (~15 to 115 µm) and different heights, occasionally short (~100 µm) lateral ridges between the longitudinal ridges; c) Knobbed surface structure of a serpent star - evenly distributed and about 10 µm in diameter; d) Microtopography of the blue mussel periostracum - 1 to 1.5 µm wide, parallel microripples which run orthogonal to the growth rings of the mussel shell[4].

and Chapter 3.4 (drag-reducing surfaces). The manufacture of such structures has been covered in Chapter 2.1. Besides these examples (which are already at a relatively advanced stage of development), there are other biomimetic structures and functionalities which will be described in the chapters which follow.

3.6.2.1 Anti-fouling surfaces

Aquatic organisms have developed diverse mechanisms to protect themselves against fouling. These include mechanical and (bio)chemical means (see Chapter 3.6.3.1) and also physical effects in the form of surface structures. The mimicking of such structures is an interesting approach because the surface topography has a major influence on bioadhesion and wettability [2].

Anti-fouling surfaces have already been described in detail in Chapter 3.5. With regards to homogenous, artificial, surface structures, it was found that macrofouling larvae have a lower affinity to surfaces whose microstructure is slightly smaller than the width of the organism (attachment point theory) [3]. The results show that the manufacture of surface structures is a way of purposefully suppressing specific macrofouling organisms. However, a universal anti-fouling coating that is effective against many groups of organisms has not yet been realised using homogenous, artificial structures (see Chapter 3.5.3.3). Surface topographies taken from nature could open up further opportunities here.

In order to look at this topic a little closer, a study of Bers and Wahl (2004) is considered here for example purposes. High resolution resin replicas of natural surface structures were produced to an accuracy of < 1µm and tested for their anti-fouling properties in field experiments. Four types of aquatic organisms were used as examples: the edible crab *(Cancer pagurus)*, the blue mussel *(Mytilus edulis)*, the serpent star *(Ophiura texturata)*, and the egg case of the small-spotted catshark *(Scyliorhinus canicula)* [4]. Figure 3.63 shows images of the different surfaces and highlights the diversity of structures which exist in nature.

The results of this study showed that the observed anti-fouling effects are as different as the microtopographies and the colonising species. Although repellent effects on microfoulers and temporarily reduced barnacle settlement were observed for the replica of the egg case of the catshark, the surface of the blue mussel only showed barnacle reduction in the first week. The microtopography of the edible crab on macrofoulers (such as barnacles) was found to be effective for a period of three weeks. The surfaces of the serpent star were also found to have repellent effects on microfoulers. It was stated in summary that the observed effects do not suffice to guarantee effective protection against fouling. The actual organisms must therefore have other anti-fouling mechanisms at their disposal (e.g. surface chemistry) to protect themselves against fouling [4]. For transfer to technical surfaces, this means that merely structuring surfaces only has a partial role to play and that other measures are also required. Bio-chemical and chemical mechanisms will be returned to in a later chapter of this book (Chapter 3.6.3.1).

3.6.2.2 Further application fields

Gecko feet: Bio-inspired adhesion

Geckos are remarkable creatures. They are able to move on a wide range of surfaces regardless of the slope and can even move headfirst. The reason for this is that they have very effective surfaces on the undersides of their feet, consisting of multi-level hierarchical structures. Firstly there are lamellae on which there are microscopic hairs (setae). These setae then in turn split into nanoscopic spatulae (see Figure 3.64). These structures mean that there is a large contact surface between the underside of the feet and the surface. It is this fact which enables geckos to adhere to a wide range of surfaces. Studies have shown that the adhesion strength is affected by the angle of the gecko setae to the surface and is a maximum at 30° [5, 6]. Detachment from the surface takes place by changing the angle.

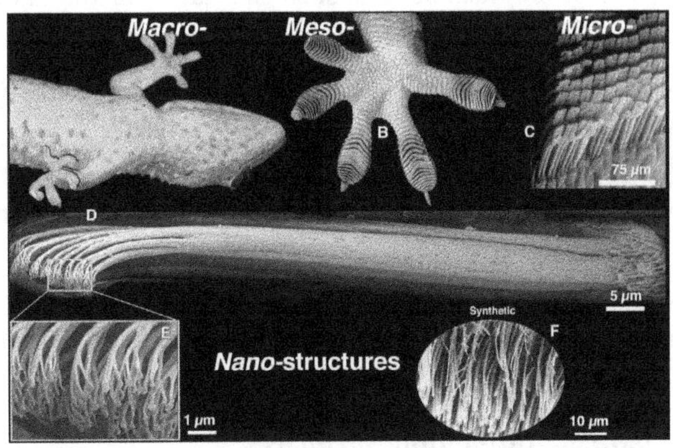

Figure 3.64: Gecko feet at different magnifications, showing the multi-level hierarchical structure with microscopic hairs (setae) and nanoscopic spatulae (with permission of Kellar Autumn) Source: http://www.kellarautumn.com/photography/Images/Pages/Geckos.html#29; visited 09-29-2010)

The complex hairy structures on the underside of the feet of geckos provide excellent dry adhesion. Two mechanisms for this have been identified: firstly van der Waals forces as the primary adhesive mechanism, and secondly capillary forces as a contributing factor [7]. How though is it possible to transfer these surface structures to technical surfaces in order, for example, to produce adhesive tapes? Various work groups have sought to answer this question. Most of the studies that have been described have produced single-level fibrillar structures, for example by using nanoindention, lithography, self-assembled layers, or carbon nanotubes. Some of these methods have already been described in this book. Problems include the bunching of the fibrilles and their stability. The studies also demonstrated that a certain degree of compliance is necessary in order to adapt to the roughness profile of the mating surface. Satisfactory longevity of the artificial structures has also not yet been demonstrated. Nevertheless, these bio-inspired structures are deemed to have huge potential. Research in this area is hence ongoing and a detailed overall insight is offered, for example, by Bhushan (2010) [5].

Optical surfaces

The diversity of optical surfaces in the animal and plant kingdoms is enormous. For example, there are various structures which induce special colourings. Silver shimmering surfaces of fish are based on crystal multilayer structures of guanine. The spectrum of colour is particularly large for butterflies. In the iridescent blue butterfly *Morpho rhetenor* the optical effect is due to structures of discrete multilayers of cuticle and air [8].

A further application of particular interest for technical surfaces is the mimicking of anti-reflective surfaces. Studies have shown that specific nanostructures reduce the reflections over broad angles or frequency ranges. This has been observed in the ommatides in the compound eyes of arthropods (e.g. moth eyes) which possess so-called nipple arrays (see Figure 3.65). This surface structuring significantly improves the photon collection efficiency [9]. Transfer of such structures to technical surfaces can lead to increased transmission and suppression of ghost images or veil glare [10].

Various methods can be used to reproduce these structures (as already described for other bio-inspired structures). However, for optical surfaces the structuring of the substrate is often paramount. Various lithographic methods including electron-beam etching, interference lithography, and colloidal lithography can be used [10]. These technologies are far removed from the coating methods described in this book and are hence not mentioned further here. Technical applications of these structures are found, for example, in optical and electro-optical devices such as solar cells and flat panel displays [10].

3.6.3 (Bio)chemical-inspired functions

Besides structure-based functions, (bio)chemical-inspired effects from nature are also being studied and transferred to technical surfaces. The chapters which follow describe various functionalities, including anti-fouling, anti-icing and self-healing systems.

There are a huge number of different (bio)chemical-inspired functions and only a few selected examples can be discussed in this book. The selected examples will give readers an initial insight into the potential of biomimetic systems for new coating technologies.

3.6.3.1 Anti-fouling surfaces

As already described briefly in Chapter 3.5.2.2, aquatic organisms have various biochemical strategies at their disposal for protecting themselves against fouling. These include the synthesis of secondary metabolites and “colonisation” with microorganisms. These and other examples are now explained in detail in relation to transfer of these systems to technical surfaces.

Figure 3.65: Examples of corneal nipple arrays as anti-reflective surfaces [9]
source: D. G. Stavenga, University of Groningen, The Netherlands

Secondary metabolites

The first biomimetic approach described here is the use of secondary metabolites on technical surfaces. Secondary metabolites are synthesised by aquatic organisms in order, amongst other things, to protect them against fouling. The classes of substances to which they belong include the terpenoids, alkaloids, fatty acids, polyphenolics, and peptides. Organisms which up until now have been identified as containing the largest amount of such substances are sponges, algae, and cnidarians (e.g. corals) [11]. A summary of secondary metabolites has been given, for example, by Fusetani (2004) [12]. The large number of substances is the reason for the large number of functions of secondary metabolites. They can, for example, influence metabolism of the fouling species or inhibit attachment, metamorphosis, and growth. Some can act as biocides and some can prevent biofilm formation by dissolving the relevant adhesives [13].

Using secondary metabolites to manufacture effective anti-fouling coatings involves a number of development steps. The first step is to identify the active substances. Sometimes the molecules are very complex, and the active components must be identified in order to enable synthetic analogs to be produced in sufficient quantity. Furthermore, substances having a wide spectrum of activity are beneficial here in order to prevent fouling by as many fouling organisms as possible.

Once substances having the aforementioned properties are available, these must be incorporated into a coating matrix, without the anti-fouling effect being lost and without adversely affecting the properties of the coating. Furthermore, the lifetime of the coating must meet the requirements for the relevant field of application (ranging from hours to weeks for medical applications, and up to years for marine applications). It quickly becomes apparent how complex a task it is to develop novel biomimetic coatings.

The mechanisms of secondary metabolites and technical implementation of the biological principles are discussed in the following in more detail using three selected classes of substances:

- polymeric 3-alkylpyridinium salts (poly-APS),
- furanones, and
- peptides.

Polymeric 3-alkylpyridinium salts (poly-APS) are alkaloids that are isolated from the Mediterranean sponge *Reniera sarai* (see Figure 3.66). Natural poly-APS can inhibit both biofilm

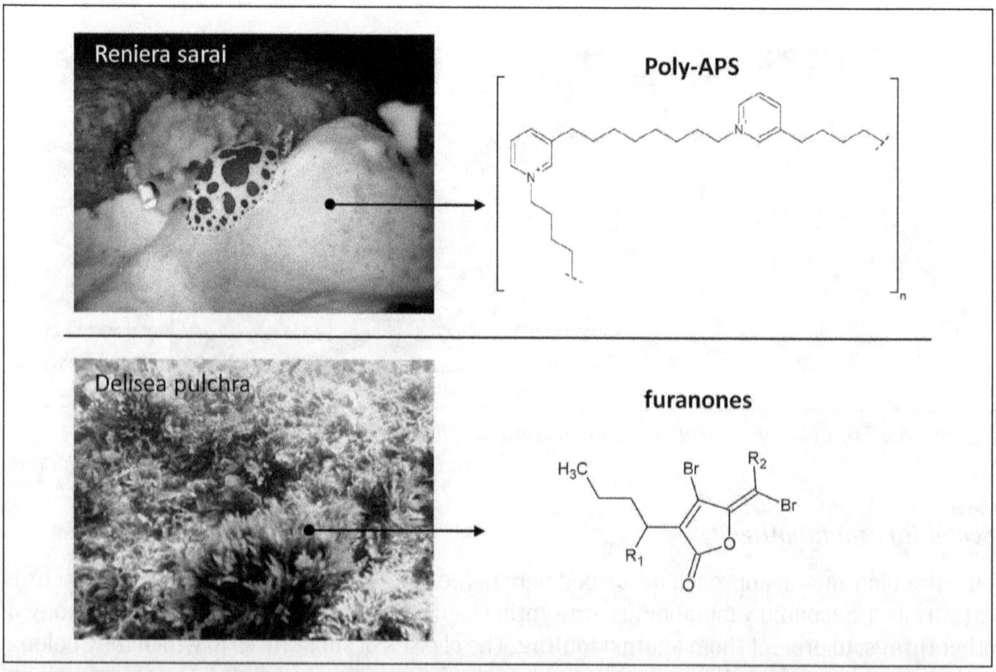

Figure 3.66: Images of organisms that synthesise secondary metabolites. Reniera sarai (sponge) (image kindly provided by Marko Gasparič, GRM Ljubljana, Slovenia) and Delisea pulchra (red algae) (image kindly provided by R. de Nys, James Cook University, Townsville, Australia) along with the synthesised products: poly-APS [15] and furanones [18].

formation (microorganisms) and the settlement of macrofoulers such as barnacle larvae. Reasons that have been put forward for this behaviour include their detergent-like properties and their ability to decrease the surface tension [14]. A study on the structure-property relationships of natural and synthetic (polymeric) 3-alkylpyridinium salts in relation to microfoulers identified the length of the alkyl chain, the presence of the charged pyridinium moiety, and the bromine atom as determining factors for the anti-fouling effects [15]. Using this knowledge, synthetic analogs having similar activity to naturally occurring poly-APS were identified. Further studies concerning the attachment of these substances to surfaces and their incorporation into coating formulations are not known to the authors of this book. Pyridinium compounds are generally known for their anti-bacterial properties (described, for example, by Scott and Gorman (1998) [16]). Various patents – without a biomimetic context – describe their use as biocidal agents. These include a patent on functionalised polysiloxanes that have biocidal functionality from the use of pyridinium compounds [17].

Another interesting class of substance is the furanones. Halogenated furanones are synthesised by the Australian red alga *Delisea pulchra* to protect itself from fouling (see Figure 3.66). They are present at the surface of the algae in special gland cells. The maximum concentration of furanones at the surface was found to be about 100 ng/cm^2 [19]. Furanones have no or only a small effect on the growth or survival of the bacteria, rather they hinder bacterial colonisation and biofilm development by acting on the bacterial signal based regulatory system. This quorom-sensing (cell to cell communication) is decisive in high-density bacterial populations as it controls, amongst other things, the reproduction process of bacteria [20, 21]. In addition to the described effect on microorganisms, furanones hinder the settlement and growth of macrofouling organisms, and they have a similar toxicity to or higher toxicity than other anti-foulants. They are, for example, an order of magnitude more effective against barnacle larvae than copper [22].

Studies aimed at transferring the anti-fouling effects of furanones to technical surfaces have been undertaken for marine and medical applications. This is reflected by the large number of patents in these areas. One patent from 2003 [23], for example, describes a range of extruded polymers incorporating furanones for marine and fresh water applications. Suitable polymers were cited as including ethylene-vinyl acetate copolymer (EVA), high-density polyethylene (HDPE), nylon, and polypropylene (PP). In field tests the manufactured materials showed excellent broad-spectrum anti-fouling activity for at least 100 days [23]. Another patent published in 2003 [24] describes surfaces for biomedical applications (although not exclusively) on which furanones are immobilised. Various bonding mechanisms and processes are described for covalently bonding furanones to different materials [24]. Finally, a patent from 2009 [25] covers coatings, including a film-forming polymer in combination with a halogenated furanone for medical devices, packing materials, and textiles. The film-forming polymers described here are in general biocompatible/biodegradable organic compounds composed, for example, of glycolides, lactides, caprolactones, and others. An advantage of using furanones is stated as being the fact that there is no risk of resistance formation as is the case, for example, with antibiotics [25].

All the various development steps have been undertaken for furanones, from the identification of the active substance to the development of effective anti-bacterial surfaces. However, no information is yet available about commercially feasible coatings. To date, no commercial furanone-based anti-bacterial/anti-fouling coatings are known to the authors of this book.

Common to all the mentioned examples is that the anti-fouling agents are formed by the macroorganisms that are to be protected. Nature also has strategies whereby organisms host other microorganisms (generally bacteria) which produce the anti-fouling agents. There are various examples of these epibiotic or endosymbiotic relationships. For example, anti-fouling extracts of epibiotic bacteria isolated from living surfaces of seaweed were formulated into water-based paints. In laboratory tests some showed effectiveness against fouling bacteria and against macrofoulers. As such, microorganisms are also therefore a promising source of natural anti-fouling compounds [26]. This was also evident in a study in which the growth of marine microorganisms on surfaces was suppressed under laboratory conditions by the use of an acrylic-based paint resin containing bacterial extracts [27]. The authors have no information about the further progress of this work.

Peptides

Anti-microbial peptides are another interesting group of secondary metabolites. Peptides are built from amino acids, with the only difference to proteins being the number of amino acids in the chain (roughly-speaking peptides contain up to about 50 amino acids and proteins more than 50). The bonding of these molecules to surfaces has already been discussed in Chapter 2.6. The chapters which follow describe selected examples of anti-fouling systems from the literature in more detail.

The majority of the anti-microbial peptides described in the literature have been studied with the aim of developing medical applications. They are seen as alternatives to the currently used antibiotics for killing the ever more resistant bacteria that are harmful to health. During the course of this work, the bonding of these substances to the surfaces of, for example, implants have been researched. Often these are cationic anti-microbial peptides having broad-spectrum activity and rapid bactericidal activity. They are synthesised by virtually all species and are numerous - hundreds have already been identified. Common to all of them is that they have a net positive charge and approximately 50 % hydrophobic residues. As a result, the molecules are present in an amphiphilic conformation and can hence interact with bacterial membranes [28]. The mechanism involves penetration of the cell membrane causing the permeability to increase and impairing the membrane function (see, for example, Bagheri et al. 2009 [29]). The effect is similar to that of the already described polymeric 3-alkylpyridinium salts (poly-APS) which have been isolated from the Mediterranean sponge *Reniera sarai.*

Known anti-microbial peptides include magainin, lactoferricin, and bactenecin. The latter (bactenecin) is a naturally occurring bovine host defense peptide and is one of the smallest anti-microbial peptides (12 amino acids). It is effective against, amongst other things, gram-positive and gram-negative bacteria [30]. Studies have also been carried out on the bonding of magainin to polymer brushes. Here, the natural anti-bacterial peptide magainin was bonded to polymer brushes based on PEG chains. The aim was to combine the repelling effect of the PEG chains and the bactericidal effect of the peptides. The covalently bonded peptides showed biocidal activity against two different strains of gram-positive bacteria [31]. Further information about this approach can be found in Chapter 2.5.4.2.

Studies on the anti-microbial activity have also been undertaken on a further group of molecules, the so-called peptoids. Both peptides and peptoids have a protein-like backbone, but differ in the localisation of the side-chains. In peptides these are bonded to the α-carbon, in peptoids to the amide nitrogen [32]. Features of peptoids are their stability, ease of synthesis, and low cytotoxicity. They are therefore interesting alternatives to peptides [33]. The immobilisation of these molecules with the aid of DOPA (dihydroxyphenylalanine) related molecules are the topic of intense ongoing research. This approach combines two bio-inspired mechanisms: the anti-microbial effect of the peptoids and the adhesion-promoting effect of DOPA. For further information about adhesion promotion by DOPA, see Chapter 2.6 and Chapter 3.6.3.4. The studies on the anti-bacterial properties of immobilised peptoids have shown that surfaces modified in this way are capable of compromising the membranes of attached bacteria. Further work is required to investigate the increased bacterial adhesion that was observed by comparing modified surfaces with bare TiO_2. It is being investigated whether this can be re-balanced by the anti-bacterial activity of the peptoids and the extent to which modifications can increase this activity [33].

Having discussed studies relating to medical applications, we will now return to an example of biologically active marine peptides. Various brominated cyclopeptides have been isolated from the marine sponge *Geodia Barretti*, with one representative being barettin (cyclo[(6-bromo-8-entryptophan)arginine]). This was incorporated together with another representative (8,9-dihydrobarettin) in a self-polishing coating and resulted in significant reduction of *M. edulis* (blue mussel) and *B. improvisus* (barnacle) settlement in field tests [34]. No specific surface bonding of the peptides was carried out in these studies. Other studies aimed at acquiring a better understanding of structure-property relationships showed that both the bromine atoms and the stereochemical properties are of key importance for the anti-fouling effect. This knowledge allowed synthetic analogs with improved activity to be manufactured [35]. Subsequent studies identified other effective dibrominated cyclopeptides [36], giving a further way of using chemical compounds following nature's example to produce technical surfaces with anti-fouling properties.

Enzymes

As was the case for secondary metabolites, there are also a huge number of enzymes and only a few selected examples can be mentioned here. For further information see, for example, the article by Kristensen et al. (2008) [37] which gives a comprehensive account of available enzymes. Enzymes are proteins that are synthesised by organisms. Enzymes catalyse biochemical reactions. In contrast to secondary metabolites, enzymes take on anti-fouling functions less often in a biological context. Nevertheless, enzymes can lend anti-fouling

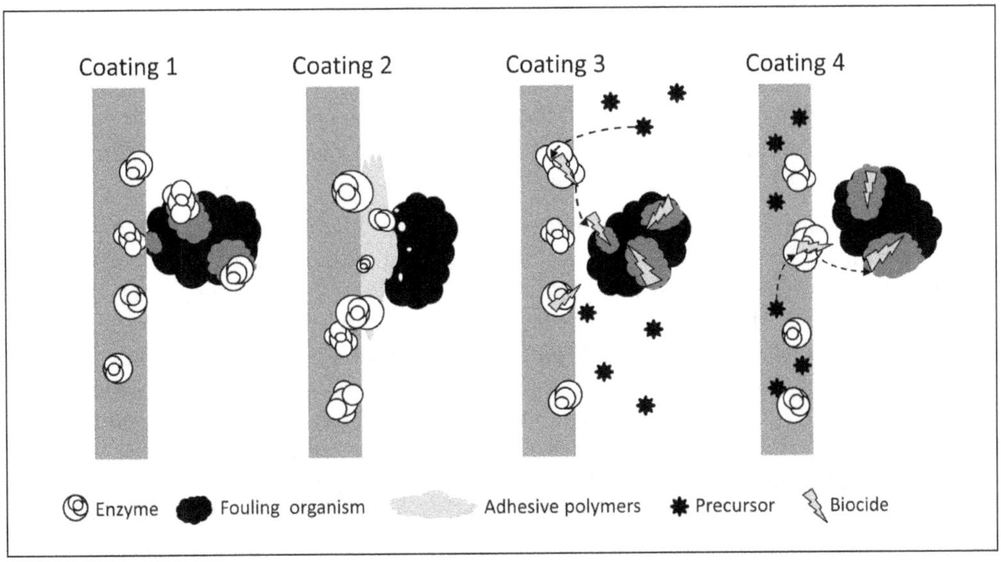

Figure 3.67: Mechanisms of enzyme-based anti-fouling coatings. 1) Direct biocidal action, 2) Adhesive degrading action, 3) Indirect action via precursors in the surroundings, 4) Indirect action via precursors in the coating *source: S.M.Olsen, Hempel A/S, Denmark and ref.* [38]

properties to technical surfaces. An example is hydrogen peroxide formation (already discussed in Chapter 3.5.5.2) via enzymatically controlled reactions.

In principle, the anti-fouling effect on technical surfaces can be achieved via four different mechanisms (according to Olsen et al. 2007) [38]:

1) Direct biocidal action of the enzyme
2) Direct adhesive degrading action of the enzyme
3) Indirect action via reaction of the enzyme with precursors in the surroundings
4) Indirect action via reaction of the enzyme with precursors in the coating

The mechanisms are schematically shown in Figure 3.67 and are explained below using examples.

Enzymes that exert a direct biocidal effect (mechanism 1) include, for example, cell wall degrading enzymes such as lysozyme, chitinase, and hyaluronidase. Chitinase catalyses the decomposition of chitin, which is an essential constituent of the barnacle exoskeleton. Although several patents are available from the 1980s and 1990s, the biocidal effects of enzymes have not been fully investigated [38]. Recent research work on latex paint film composites containing carbon nanotube - enzyme conjugates for medical applications has however been very promising. The enzyme that was employed, lysostaphin (a cell wall degrading endopeptidase), was covalently bonded to MWNTs and showed bactericidal effects without leaching form the paint matrix [39]. Details about the bonding have already been given in Chapter 2.6.

For adhesive degrading enzymes (mechanism 2), more extensive information is available. Some overlap with biocidal mechanisms (mechanism 1) cannot be excluded here. Fouling organisms secrete adhesives to anchor themselves to surfaces. These adhesives are based on different classes of materials such as glycoproteins, carbohydrates, and proteins depending on the type of organism. This means that different classes of enzymes must be used in a coating (a protein-degrading enzyme and a polysaccharide-degrading enzyme) in order for the coating to have a wide spectrum of activity against fouling organisms. In addition, the enzymes must be active immediately the fouling organism approaches the surface so that adhesive that has not yet cured is hydrolysed; effective enzymatic degradation of cured adhesives, especially under the exclusion of water that is essential for the enzymatic reaction, is considered to be very unlikely. If these conditions are fulfilled, adhesive degrading enzymes can weaken the bond strength and hence foulants can be more easily removed by hydrodynamic forces [38].

Various studies have been conducted on this mechanism. Different classes of enzymes were investigated for their anti-fouling activity [40]. Proteases were found to have much higher activity than the glycolytic and lipolytic enzymes that were tested. The observed effects were not due to toxic effects (the fouling organisms were not killed), but were rather due to prevention of colonisation. Proteases belonging to deep-sea bacterial isolates of the genus *Pseudoalteromonas sp.* were found to be particularly effective. After incorporation into water-based paint, these enzymes gave the best results in field tests. With regards to the enzyme stability under the field conditions (seawater), time periods of 14 days were indicated. The optimal temperature was stated as being 30 °C and the optimum pH was indicated as 8, namely close to conditions in tropical seawater [40]. These parameters largely determine the enzyme activity and hence the fields of application for these biomimetic systems.

Regarding the third mechanism (indirect action by reaction of the enzyme with precursors in the surroundings), Olsen et al. (2007) described a haloperoxidase catalysed reaction in which the enzyme has an indirect anti-fouling effect via the reaction of halide ions in seawater with hydrogen peroxide to form hypohalogenic acid (e.g. HOBr). The latter has an oxidising effect. They add, however, that the concentration of hydrogen peroxide in seawater is too low to produce an adequate amount of hypohalogenic acid [38].

Enzyme-mediated hydrogen peroxide release can be assigned to the fourth mechanism: indirect action via reaction of the enzyme with precursors in the coating. In the described reaction the anti-fouling agent (hydrogen peroxide) is formed in two steps (firstly the conversion of starch into glucose catalysed by glucoamylase and secondly the hexose oxidase catalysed formation of hydrogen peroxide). The release of the anti-fouling agent can hence be controlled via enzymes [38]. Other studies have been undertaken on this system, and these will be mentioned later in this chapter.

Enzyme-based anti-fouling coatings have been researched for almost 30 years. Many patents have been registered and these have been summarised by, amongst others, Olsen et al. (2007) [38]. The authors are, however, unaware of any enzyme-based products currently in the marketplace. The main hurdle which has to be overcome (as is also the case for coatings containing secondary metabolites) is guaranteeing the anti-fouling effect after incorporation into the coating matrix. In addition, the optimal enzyme activity is limited to a relatively narrow region of pH, temperature, and salinity. Furthermore, the stability of the enzyme structure often does not meet the required duration for surface coatings [38].

Regarding the hydrogen peroxide releasing enzyme system, various studies have already been conducted to solve the mentioned shortcomings. For examples, studies showed that physical encapsulation of the enzymes gives improved enzyme stability and concomitant prolonged lifetime of the anti-fouling coating [41] (Chapter 2.6). This involved modifying the hydrogen peroxide producing enzyme system. The anchoring of the hexose oxidase enzyme on silica via polycation-templated silica co-precipitation produced particle sizes of 0.7 μm to 17 μm (95 % of the colloidal volume). This enabled the activity and stability of HOX to be considerably improved. (Similar procedures were unnecessary for glucoamylase because tests showed that here the starches acted as anchors.) The coating system formulated with this modified enzyme-complex was immersed in artificial seawater under static conditions and showed hydrogen peroxide release over 70 days [41]. Further studies on the anti-fouling properties under field conditions in the North Sea gave positive results after immersion for 97 days [42]. However, no positive effects for this H_2O_2-releasing coating could be achieved in warmer climates (offshore near Singapore and Barcelona) [43]). Further research is still required.

Further measures

One example of the utilisation of microorganism-related non-toxic compounds is tetraether lipids from the archaea bacterium *Thermoplasma acidophilum*. These bacteria are very resistant to strong acids (pH=2) and high temperatures. This is due to the structure of the membrane surrounding the organisms. It consists of a tetraether core with two methyl-branched, saturated C_{40} hydrocarbon chains strongly bonded to two glycerol moieties via ether linkages. This chemical structure leads to very good resistance to hydrolytic, oxidative, and other (bio)chemical attack [44]. But what uses could this have for technical surfaces? Of interest here are to a lesser extent the biocidal effects and to a greater extent the physico-chemical properties of the surfaces modified with tetraether lipids. In two studies on the covalent bonding of tetraether lipids to silicon materials [45] and polyurethanes [44] (in both cases for medical devices), the initial bacterial adhesion was significantly reduced within 24 hours. These (repelling) anti-fouling effects were no longer detectable after the first colonisation by bacteria; biofilm growth and colonisation were hence not influenced by these non-toxic, biocompatible coatings [44, 45].

Another biologically inspired approach is the replacement of conventional pigments (Cu_2O and ZnO) by a starch-enzyme complex to realise the self-polishing effect for marine applications [46]. Due to the glucoamylase catalysed degradation of starches to water-soluble glucose, polishing rates of 7 to 10 µm per month were achieved. The most suitable here was starch with small water soluble content (in this case corn starch) because here the swelling related to the water uptake could be reduced and the mechanical stability improved [46]. This interesting principle avoids the use of metal-containing pigments and hence represents an environmentally-friendly alternative. As the tested coatings contained no biocidal components and the availability of nutrients at the surface (dissolved glucose) could have an undesired positive effect on fouling organisms, there is also a need here for further development work.

3.6.3.2 Biomimetic anti-icing surfaces

The bonding of proteins to surfaces is one of the main themes in the development of (bio) chemical-inspired coatings. Chapter 2.6 has already described in detail various methods for this. One important function of such coatings is the prevention or reduction of ice formation on surfaces. Chapter 3.2 has already given information about this. One major approach has, however, not yet been described, namely the use of so-called anti-freeze proteins (AFPs). This biomimetic approach aims to transfer anti-icing mechanisms that occur in nature to technical surfaces.

Since the discovery of AFPs in fish in 1969 [47], these natural anti-freeze agents have been found in a variety of organisms such as amphibians, insects, and plants. They allow these organisms to survive in cold regions of the world. (An overview of the structural diversity of AFPs is, for example, given by Grunwald et al. (2009) [48]). Different to the freezing point suppressors (see Chapter 3.2) with their colligative properties (e.g. dissolved salts), the anti-icing effects are due to the constitutive properties of the anti-freeze proteins. This means that they reduce the freezing point due to their molecular structure. The freezing point reduction can be as much as a thousand-fold greater than would be expected from the molar concentration alone (the colligative effect) [49]. Another special aspect is that the AFPs lower the freezing point but not the melting point (see Figure 3.68). This temperature

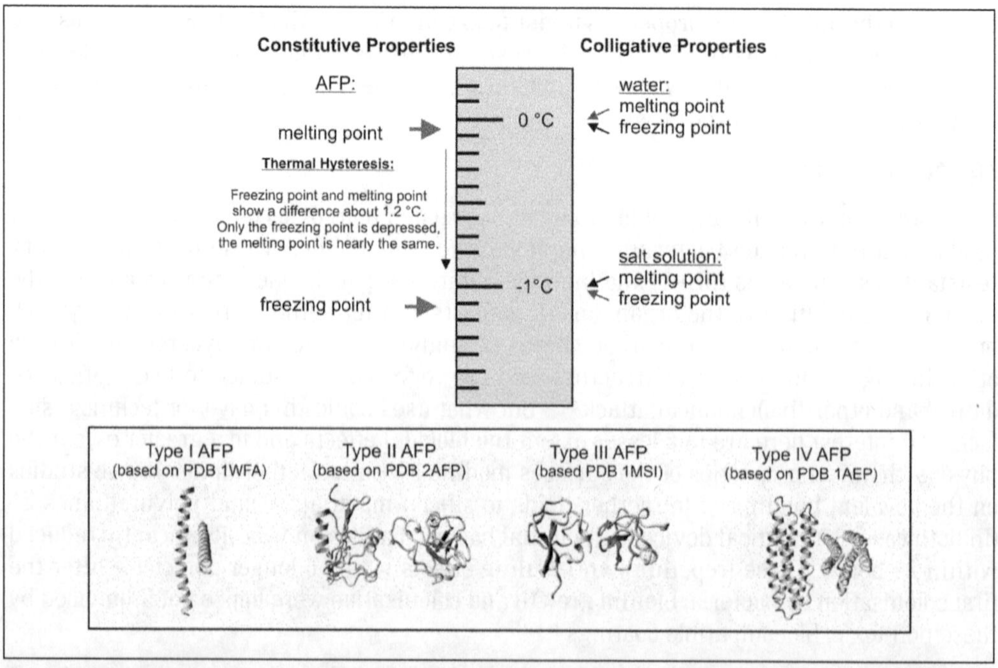

Figure 3.68: Thermal hysteresis of anti-freeze proteins (AFP) with comparison of constitutive and colligative properties. Furthermore, a selection of 3-D models of AFPs is shown derived on the basis of structural data of corresponding proteins from the RCSB protein database. Modelling was performed with the software YASARA *Source: I. Grunwald, Fraunhofer IFAM, Bremen*

difference is termed thermal hysteresis and anti-freeze proteins are hence often referred to in the literature as thermal hysteresis proteins [49].

The complex anti-icing mechanisms of AFPs have been the topic of many research studies but are still not fully understood. Even an introduction to this topic is outside the scope of this book, but this much is known: certain regions of the AFPs bind ice crystals and change the morphology of the ice crystals in such a way that continuation of the crystallisation to form a closed ice layer is suppressed and hence the freezing point is lowered (for further details see, for example, ref. [50-52]).

In order to be able to transfer this function of the AFPs to technical surfaces, the active sequences of the proteins must first of all be localised and synthesised via solid phase peptide synthesis (SPPS). Then, the synthetic peptide sequences must be bonded to suitable surfaces such that their functionality is preserved. Various processes are available for this, as explained in Chapter 2.6. The most recent research work on this biomimetic approach has shown that it is possible to bond anti-freeze proteins to paint surfaces with the help of linker molecules and that the activity of the proteins is maintained. In addition, tests have been carried out in the ice chamber described in Chapter 3.2 which showed that ice formation on such surfaces is reduced (see Figure 3.69). Further research studies on biomimetic anti-icing surfaces are ongoing.

3.6.3.3 Coatings containing living organisms

A further biomimetic approach is the incorporation of living organisms into coatings. Although the use of non-bonded (free) microorganisms (or enzymes) in biotechnology is widespread (ranging from beer, bread and cheese production to wastewater treatment), the use of living organisms immobilised in coatings is still in its infancy. So what benefits could these microorganisms have for surfaces? To answer this, let's consider first of all the example of anti-fouling coatings. If living cells can be effectively incorporated, the service life of the coating can be prolonged due to the continuous production of anti-fouling agents [38].

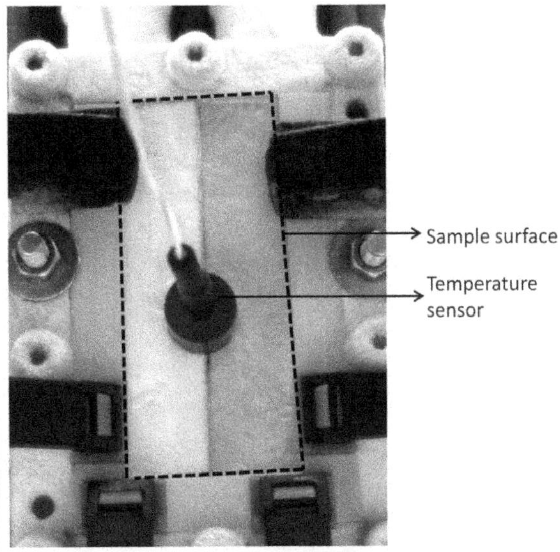

Figure 3.69: Reduced ice formation due to the bonding of AFP sequences to paint surfaces. Left: Sample without AFPs having a closed ice film thickness of 500 to 550 μm; Right: Sample with AFPs with non-uniform ice formation and a maximum ice film thickness of 350 μm. Test conditions: Air temperature 1 °C, relative humidity: 88 %, wind speed 9 m/sec, substrate temperature -1.5 °C, test duration 1 h

source: Fraunhofer IFAM

As one can imagine, this task is very complex and requires lengthy research. First of all suitable bacteria are required that are able to produce sufficiently effective anti-fouling agents and are able to withstand entrapment in the coating. For example, *Pseudoalteromonas tunicata*, the bacteria strain isolated from adult tunicates from waters off the Swedish west coast, suppresses the settlement of common biofouling organisms via the production of various active compounds [53]. Incorporation of this bacteria strain in a polyacrylamide-based hydrogel is described in literature [54]. Here the bacterial viability was dependent on the monomer concentration of the acrylamide. In this initial study, the presence of living bacteria and their anti-fouling effect against barnacle larvae was determined over a period of only a few days [54]. In a further study, anti-fouling activity for up to two weeks was measured for a Dupont polyvinyl alcohol (PVOH) 10 % gel [55]. Other slight improvements were achieved by using beads in the gels, so enlarging the surface for attachment of the bacterial cells. In contrast, no success was booked by attempts to minimise the stress of incorporation experienced by the bacteria (pH changes, dehydration, physical stress). The use of more stress-resistant bacteria strains such as *Eschericha coli* in combination with particles gave a maximum lifetime of 80 days [55].

Anti-fouling coatings based on classical paint chemistry are covered in a patent from 1999 which describes, amongst other things, epoxy and polyurethane coating materials in combination with microorganisms and/or hydrolytic enzymes [56]. Scientific proof that the demonstrated anti-fouling properties were due to the presence of living microorganisms in the cured paint layers is however not available to the authors.

Apart from the functions which the so-called living paints have to take on, these coatings must guarantee the survival of the organisms and the continuous production and transport of active agents. This requires continuous metabolic activity, transport of nutrients, and removal of metabolic waste products. In addition, the coatings must fulfil the general requirements for their technical application [55]. This also applies for the sol-gel systems described in Chapter 2.4, for which various literature sources describe the incorporation of living cells. It has been studied, for example, how to minimise the negative effect of the alcohols produced as byproducts by the sol-gel process. The characteristic shrinkage of sol-gel silica matrices also has negative effects for the viability of entrapped cells and this can

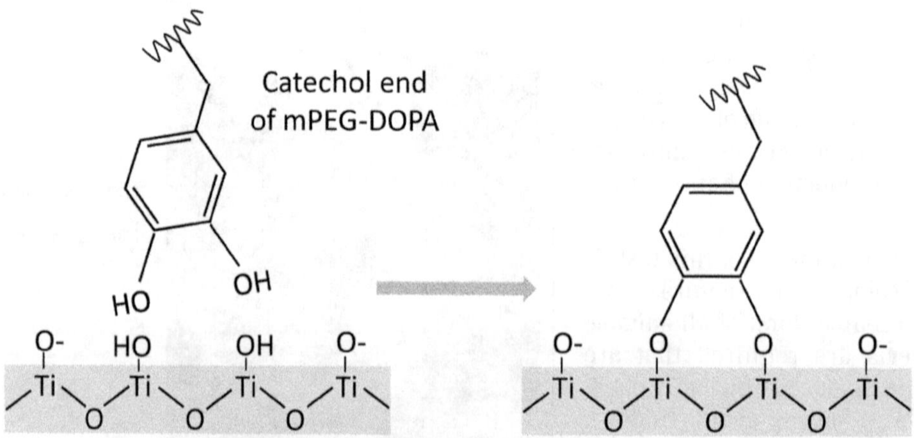

Figure 3.70 Chemical structure of mussel adhesive inspired mPEG-DOPA3 used for reducing marine fouling [62]

be suppressed by choice of the organic-inorganic hybrid matrix (see, for example, ref. [57, 58]). Potential applications of these sol-gels containing microorganisms include, for example, the biodegradation of phenols and polychlorinated biphenyls (PCBs) [59] and application as viable cellular sensors [60].

Latex-based coatings have also been the focus of other studies. The incorporation of living cells here should be an inexpensive solution for stabilising microorganisms, so improving their viability und activity compared to microorganisms in suspensions [61]. Here, latex coatings containing ~50 % microorganisms by volume are described. The entrapped organisms are surrounded by nanopores, which are vital for the microbial viability and coating reactivity. Besides model systems which allow oxidative processes (e.g. for the manufacture of vitamin C), photoreactive coatings for light adsorption and hydrogen production are also described (for further details and examples see Flickinger et al. (2007) [61]).

The use of living organisms might prove to be a smart solution for biotechnology challenges, because biological processes can be directly integrated here into coating concepts. It remains to be seen whether the hurdles regarding survival during the film forming processes can be overcome and whether the duration of the reactivity can be optimised, and hence expectations fulfilled.

3.6.3.4 Further application fields

Self-healing surfaces

The principle of healing (or repair from a technical viewpoint) is well established in the animal and plant kingdoms. Healing processes are generally very complex and are based on the growth and reproduction of living cells, namely processes which cannot be transferred to technical surfaces. However, chemical-based mechanisms have successfully been developed to "heal" damaged polymer surfaces. The mechanisms cover, for example, the microencapsulation of crosslinkable healing agents (Chapter 3.3.2.1) and the re-formation of broken chemical bonds using UV light (Chapter 3.3.2.3). The biological principle of self-healing was achieved here using alternative, chemical-inspired processes.

Adhesion promotion with DOPA

A further interesting biomimetic research field is the development of adhesives following the example of nature. Key research areas here are clarification of the mechanisms involved and transfer to technical surfaces. The proteins produced by the blue mussel *Mytilus edulis* and especially DOPA (dihydroxyphenylalanine) containing compounds are an example. These are, for example, being bonded onto the polyethylene glycol (PEG) chains described in Chapter 3.5.3.4 in order to improve the adhesion properties of these anti-fouling compounds and hence to prolong the durability of the coating. The underlying chemical processes have been explained in detail in Chapter 2.6.

At this point referral is solely made to the study of Statz et al. (2006) [62] which showed that the polymer shown in Figure 3.70, comprising methoxy-terminated poly(ethylene glycol) (mPEG) and the amino acid DOPA, can be used to manufacture coatings having anti-fouling and also foul-release properties. The adhesive nature of the DOPA appears here to provide robust anchorage for the PEG chains [62].

Sandfish – minimal solid friction

This last example of bio-inspired surfaces once again highlights the steps that have to be taken to develop a biomimetic product: Observations of sandfish *(Scincus scincus)* which live in the desert (see Figure 3.71) showed that these animals move via a type of swimming motion below the sand surface. First studies on the surface of sandfish showed that their skin has a very low friction coefficient for sand and other materials and has very good abrasion resistance [63]. What though is the reason for these unusual properties? It was surmised that the low friction was due to special surface structures which allow sliding through the sand. Studies by Baumgartner et al. (2007) [64] could however not confirm this. Moreover, they found that the properties were due to a homogenous, pure organic material. The scales which cover the sandfish are made of ß-keratin which is additionally glycosylated. This gives them their extraordinary surface properties [64].

Figure 3.71: Image of Scincus scincus

source: Werner Baumgartner, RWTH-Aachen, Germany

The authors have little information about the status of research on technical application of these relative new biomimetic surface functions. Baumgartner et al. (2007) successfully reconstructed the scales by deposition of the (original) glycoproteins and suggested applications for, amongst other things, touch screens and mechanical construction elements such as bearings and gearwheels [64].

3.6.4 Conclusions

Biomimetic surfaces are a fascinating field of research. The realisation of biological functions on technical surfaces offers huge promise (if successfully implemented). In some areas, for example for drag-reducing and dirt-repelling surfaces, the development work is very advanced. In other areas that have been described here, development work is still at the early stages: one example is the development of biomimetic anti-fouling coatings. Despite the effort that has been put into this area, there are as yet no market-ready products available – in particular for the lucrative area of marine coatings. It is also not clear whether other technical developments are yet ready for implementation.

The selection of biomimetic surface concepts that have been presented here give readers an initial insight into future developments and their potential. It should now be also evident that scientific work does not necessarily result in market-ready products. Therefore, when making decisions about new research projects, the feasibility (including probability of success), commercial cost and benefit, and risk of failure always have to be weighed up.

3.6.5 Literature

[1] TAB – Büro für Technikfolgen-Abschätzung beim Deutschen Bundestag; *Oertel, D., Grunwald, A.*, Potenziale und Anwendungsperspektiven der Bionik, Arbeisbericht Nr. 108, 2006, http://www.tab-beim-bundestag.de/de/pdf/publikationen/berichte/TAB-Arbeitsbericht-ab108.pdf

[2] *Sriyutha Murthy, P., Venugopalan, V.P., Nair, K.V.K., Subramoniam, T.*, Larval Settlement and Surfaces: Implications in development of Antifouling Strategies, In: *Flemming, H.C., Sriyutha Murthy, P., Venkatesan, R., Cooksey, K.E.* (Eds.), Marine and Industrial Biofouling, Springer-Verlag Berlin Heidelberg, 2008, pp.233-263

[3] *Scardino, A.J., Guenther, J., deNys, R.*, Attachment point theory revisited: the fouling response to a microtextured matrix, Biofouling, 24 (1), 2008, 45-53

[4] *Bers, A.V., Wahl, M.*, The Influence of Natural Surface Microtopographies on Fouling, Biofouling, 20 (1), 2004, 43-51

[5] *Bhushan, B.*, Gecko Feet: Natural Hairy Attachment Systems for Smart Adhesion, In: Springer Handbook of Nanotechnology, 3rd edition, Springer-Verlag Heidelberg, Dordrecht, London, New York, 2010, pp. 1553-1597, http://www.springerlink.com/content/n182217k133j6237/

[6] *Autumn, K.*, Gecko Adhesion: Structure, Function, and Applications, MRS Bulletin, 32, 2007, 1-6

[7] *Autumn, K., Sitti, M., Liang, Y.A., Peattie, A.M., Hansen, W.R., Sponberg, S., Kenny, T.W., Fearing, R., Israelachvili, J.N., Full, R.J.*, Evidence for van der Waals adhesion in gecko setae, PNAS, 99 (19), 2002, 12252-12256

[8] *Vukusic, P., Sambles, J.R.*, Photonic structures in biology, Nature, 424, 2003, 852-855

[9] *Stavenga, D.G., Foletti, S., Palasantzas, G., Arikawa, K.*, Light on the moth-eye corneal nipple array of butterflies, Proc. R., Soc. B, 273, 2006, 661-667

[10] *Li, Y., Zhang, J., Yang, B.*, Antireflective surfaces based on biomimetic nanopillared arrays, Nano Today, 5, 2010, 117-127

[11] *Chambers, L.D., Stokes, K.R., Walsh, F.C., Wood, R.J.K.*, Modern approaches to marine antifouling coatings, Surface & Coatings Technology, 201, 2006, 3642-3652

[12] *Fusetani, N.*, Biofouling and antifouling, Nat. Prod. Rep., 21, 2004, 94-104

[13] *Yebra, D.M., Kiil, S., Dam-Johansen, K.*, Antifouling technology – past, present and future steps towards efficient and environmentally friendly antifouling coatings, Progress in Organic Coatings, 50, 2004, 75-104

[14] *Faimali, M., Garaventa, F., Mancini, I., Sicurelli, A., Guella, G., Piazza, V., Greco, G.*, Antisettlement activity of synthetic analogues of polymeric 3-alkylpyridinium salts isolated from the sponge *Reniera sarai*, Biofouling, 21(1), 2005, 49-57

[15] *Chelossi, E., Mancini, I., Sep i, K., Turk, T., Faimali, M.*, Comparative antibacterial activity of polymeric 3-alkylpyridinium salts isolated from the Mediterranean sponge *Reniera sarai* and their synthetic analogues, Biomolecular Engineering, 23, 2006, 317-323

[16] *Scott, E.M., Gorman, S.P.*, Chemical disinfectants, antiseptics and preservatives, In: *Hugo, W.B., Russel, A.D.*, Pharmaceutical Microbiology, Blackwell Science, Oxford, 1998, pp. 201-228

[17] *Boudjouk, P., Thomas*, J., Polysiloxanes with anti-fouling activity, Patent US 2009/0018276A1, 2009

[18] *Steinberg, P.D., Schneider, R., Kjelleberg, S.*, Chemical defenses of seaweeds against microbial colonization, Biodegradation, 8, 1997, 211-220

[19] *Dworjanyn, S.A., De Nys, R., Steinberg, P.D.*, Localisation and surface quantification of secondary metabolites in the red alga Delisea pulchra, Marine Biology, 133, 1999, 727-736

[20] *Steinberg, P.D., De Nys, R., Kjelleberg, S.*, Chemical Cues for Surfaces Colonization, Journal of Chemical Ecology, 28(10), 2002, 1935-1951

[21] *De Nys, R., Givskov, M., Kumar, N., Kjelleberg, S., Steinberg, P.D.*, Furanones, Prog Mol Subcell Biol, 42, 2006, 55-86

[22] *De Nys, R., Steinberg, P.D., Willemsen, P., Dworjanyn, S.A., Gabelish, C.L., King, R.J.*, Broad Spectrum Effects of Secondary Metabolites from the Red Alga *Delisea Pulchra* in Antifouling Assays, Biofouling, 8, 1995, 259-271

[23] *Christie, G.B.Y., Christov, V., De Nys, P.C., Steinberg, P., Hodson, S.*, Antifouling Polymers, Patent US 6,635,692 B1, 2003

[24] *Read, R.W., Kumaar, N., Wilcox, M., Zhu, H., Griesser, H., Muir, B., Thissen, H., Hughes, T.*, Antimicrobial coatings, Patent US 2003/0224032 A1, 2003

[25] *Stopek, J.B., Hotter, J., Tsai, S.*, Antimicrobial coatings, Patent US 2009/0182337 A1, 2009

[26] *Burgess, J.G., Boyd, K.G., Armstrong, K.G., Jiang, Z., Yan, L., Berggren, M., May, U., Pisacane, T., Granmo, A., Adams, D.R.*, The Development of a Marine Natural Product-based Antifouling Paint, Biofouling, 19, 2003, 197-205

[27] *Armstrong, E., Boyd, K.G., Pisacane, A., Peppiatt, J.*, Marine microbial natural products in antifouling coatings, Biofouling, 16, 2000, 215-224

[28] *Marr, A.K., Gooderham, W.J., Hancock, R.E.W.*, Antibacterial peptides for therapeutic use: obstacles and realistic outlook, Current Opinion in Pharmacology, 6, 2006, 468-472

[29] *Bagheri, M., Beyermann, M., Dathe, M.*, Immobilization of Surface-Bound Cationic Antimicrobial Peptides with No Influence upon the Activity Spectrum, Antimicrobial Agents and Chemotherapy, 53 (3), 2009, 1132-1141

[30] *Hilpert, K., Elliott, M.R., Volkmer-Engert, R., Henklein, P., Donini, O., Zhou, Q., Winkler, D.F.H., Hancock, R.E.W.*, Sequence Requirements and an Optimization Strategy for Short Antimicrobial Peptides, Chemistry & Biology, 13, 2006, 1101-1107

[31] *Glinel, K., Jonas, A.M., Jouenne, T., Leprince, J., Galas, L., Huck, W.T.S.*, Antibacterial and Antifouling Polymer Brushes Incorporating Antimicrobial Peptide, Bioconjugate Chem., 20. 2009, 71-77

[32] *Statz, A.R., Meagher, R.J., Barron, A.E., Messersmith, P.B.*, New Peptidomimetic Polymers for Antifouling Surfaces, J. Am. Chem. Soc., 127, 2005, 7972-7973

[33] *Statz, A.R., Park, J.P., Chongsiriwatana, N.P., Barron, A.E., Messersmith, P.B.*, Surface-immobilised antimicrobial peptoids, Biofouling, 24 (6), 2008, 439-448

[34] *Sjögren, M., Dahlström, M., Göransson, U., Jonsson, P.R., Bohlin, L.*, Recruitment in the Field of *Balanus improvises* and *Mytilus edulis* in Response to the Antifouling Cyclopeptides Barettin and 8,9-Dihydrobarettin from the Marine Sponge *Geodia baretti*, Biofouling, 20 (6), 2004, 291-297

[35] *Sjögren, M., Johnson, A.L., Hedner, E., Dahlström, M., Göransson, U., Shirani, H., Bergman, J., Jonsson, P.R., Bohlin, L.*, Antifouling activity of synthesized peptid analogs of the sponge metabolite barettin, Peptides, 27, 2006, 2058-2064

[36] *Hedner, E., Sjögren, M., Hodzic, S., Andersson, R., Göransson, U., Jonsson, P.R., Bohlin, L.*, Antifouling Activity of a Dibrominated Cyclopeptide from the Marine Sponge *Geodia baretti*, J. Nat. Prod., 71, 2008, 330-333

[37] *Kristensen, J.B., Meyer, R.L., Laursen, B.S., Shipovkov, S., Besenbacher, F., Poulsen, C.H.*, Antifouling enzymes and the biochemistry of marine settlement, Biotechnology Advances, 26, 2008, 471-481

[38] *Olsen, S.M., Pedersen, L.T., Laursen, M.H., Kiil, S., Dam-Jonhansen, K.*, Enzyme-based antifouling coatings: a review, Biofouling, 23 (5/6), 2007, 269-383

[39] *Pangule, R.C., Brooks, S.J., Dinu, C.Z., Bale, S.S., Salmon, S.L., Zhu, G., Metzger, D.W., Kane, R.S., Dordick, J.S.*, Antistaphylococcal Nanocomposite Films Based on Enzyme-Nanotube Conjugates, ACSNano, Vol.4 (7), 2010, 3993-4000

[40] *Dobretsov, S., Xiong, H., Xu, Y., Levin, L.A., Qian, P.-Y.*, Novel Antifoulants: Inhibition of Larval Attachment by Proteases, Marine Biotechnology, 9, 2007, 388-397

[41] *Kristensen, J.B., Meyer, R.L. Poulsen, C.H., Kragh, K.M., Besenbacher, F., Laursen, B.S.*, Biomimetic silica encapsulation of enzymes for replacement of biocides in antifouling coatings, Green Chem. 12, 2010, 387-394

[42] *Kristensen, J.B., Olsen, S.M., Laursen, B.S., Kragh, K.M., Poulsen, C.H., Besenbacher, F., Meyer, R.L.*, Enzymatic generation of hydrogen peroxide shows promising antifouling effect, Biofouling, 26 (2), 2010, 141-153

[43] *Olsen, S.M., Kristensen, J.B., Laursen, B.S., Pedersen, L.T., Dam-Johansen, K., Kiil, S.*, Antifouling effect of hydrogen peroxide release from enzymatic marine coatings: Exposure testing under equatorial and Mediterranean conditions, Progress in organic coatings, 2010, in Press

[44] *Sateesh, A., Vogel, J., Dayss, E., Fricke, B., Dölling, K., Rothe, U.*, Surface modifications of medical grade polyurethane by cyanurchloride-activated tetraether lipid (a new approach for bacterial antiadhesion), Wiley Periodicals, Inc. J Biomed Mater Res, 94A, 2008, 672-681

[45] *Frant, M., Stenstad, P., Johnsen, H., Dölling, K., Rothe, U., Schmid, R., Liefeith, K.*, Anti-infective surfaces based on tetraether lipids for peritoneal dialysis catheter systems, Mat.-wiss. u. Werkstofftech., 37(6), 2006, 538-545

[46] *Olsen, S.M., Pedersen, L.T., Dam-Johansen, K., Kristensen, J.B., Kiil, S.*, Replacement of traditional seawater-soluble pigments by starch and hydrolytic enzymes in polishing antifouling coatings, J. Coat. Technol. Res., 7(3), 2009, 355-363

[47] *DeVries, A.L., Wohlschlag, D.E.*, Freezing Resistance in Some Antiarctic Fishes, Science7, Vol. 163 (3871), pp. 1073-1075

[48] *Grunwald, I., Rischka, K., Kast, A.M., Scheibel, T., Bargel, H.*, Mimicking biopolymers on a molecular scale: nano(bio)technology based on engineered proteins, Phil. Trans. R. Soc. A, 367, 2009, 1727-1747

[49] *Barrett, J.*, Thermal hysteresis proteins, The International Journal of Biochemistry & Cell Biology, 33, 2001, 105-117

[50] *Madura, J.D., Baran, K., Wierzbicki*, A., Molecular recognition and binding of thermal hysteresis proteins to ice, J. Mol. Recognit., 13, 2000, 101-113

[51] *Jia, Z., Davies, P.L.*, Antifreeze proteins: an unusual receptor-ligand interaction, trends in Biochemical Sciences, 27(2), 2002, 101-106

[52] *Kristiansen, E., Zachariassen, K.E.*, The mechanism by which fish antifreeze proteins cause thermal hysteresis, Cryobiology, 51(3), 2005, 262-280

[53] *Holmström, C., Egan, S., Franks, A., McCloy,S., Kjelleberg, S.*, Antifouling activities expressed by marine surface associated *Pseudoalteromonas* species, FEMS Microbiology Ecology, 41, 2002, 47-58

[54] *Gatenholm, P., Holmström, C., Maki, J.S., Kjelleberg, S.*, Toward biological antifouling surface coatings: Marine bacteria immobilized in hydrogel inhibit barnacle larvae, Biofouling, 8(4), 1995, 293-301

[55] *Steinberg, P. Christov, V., Christie, G., Kjelleberg, S.*, Bacteria immobilised in Gels: Improved methodologies for antifouling and biocontrol applications, Biofouling, 15, 2000, 109-117

[56] *Selvig, T.A., Leavitt, I., Powers, W.P.*, Marine antifouling methods and compositions, Patent 5,919,689, 1999

[57] *Ferrer, M.L., Yuste, L., Rojo, F., del Monte, F.*, Biocompatible Sol-Gel Route for Encapsulation of Living Bacteria in Organically Modified Silica Matrixes, Chem. Mater., 15, 2003, 3614-3618

[58] *Finnie, K.S., Bartlett, J.R., Woolfrey, J.L.*, Encapsulation of sulfate-reducing bacteria in a silica host, J. Mater. Chem., 10, 2000, 1099-1101

[59] *Brányik, T., Kuncova, G., Páca, J., Demnerová, K.*, Encapsulation of Microbial Cells into Sol-Gel, Journal of Sol-Gel Science and Technology, 13, 1998, 283-287

[60] *Avnir, D., Coradin, T., Lev, O., Livage, J.*, Recent bio-applications of sol-gel materials, J. Mater. Chem., 16, 2006, 1013-1030

[61] *Flickinger M.C., Schottel, J.L., Bond, D.R., Aksan, A., Scriven, L.E.*, Painting and Printing Living Bacteria: Engineering Nanoporous Biocatalytic Coatings to Preserve Microbial Viability and Intensify Reactivity, Biotechnol. Prog., 23, 2007, 2-17

[62] *Statz, A., Finlay, J., Dalsin, J., Callow, M., Callow, J.A., Messersmith, P.B.*, Algal antifouling and fouling-release properties of metal surfaces coated with a polymer inspired by marine mussels, Biofouling, 22(6), 2006, 391-399

[63] *Rechenberger, I., El Khyari, A.R.*, The Sandfish of the Sahara – A Model for Friction and Wear Reduction, http://www.bionik.tu-berlin.de/institut/safiengl.htm (visited August 2010)

[64] *Baumgartner, W., Saxe, F., Weth, A., Hafas, D., Sigumonrong, D., Emmerlich, J., Singheiser, M., Böhme, W., Schmeider, J.M.*, The Sandfish's Skin: Morphology, Chemistry and Reconstruction, Journal of Bionic Engineering, 4, 2007, 1-9

4 Outlook

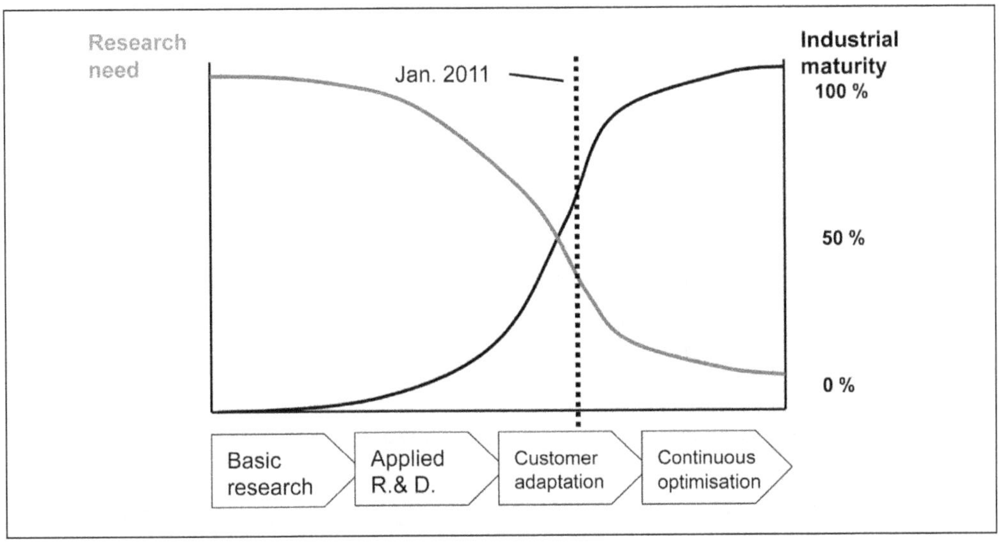

Figure 4.1: Example of a functional coating at an advanced stage of industrial maturity: photocatalytic, self-cleaning coatings

The foregoing chapters have described a variety of examples of approaches that are currently being employed to develop functional coating systems. We have also given examples of functional coatings that are in a variety of stages of development.

Photocatalytic, self-cleaning coatings are, for example, at an advanced stage of development (see Chapter 3.1 and Figure 4.1). Such coatings are already commercially available for certain applications. The R&D need here is for optimisation of the coatings for specific substrates and applications (e.g. for inside applications).

For products with this level of maturity the need for fundamental R&D work is no longer very high. However, the full potential of this technology for many applications still has to be identified and commercially realised by carrying out customised application-oriented R&D work. The R&D effort here is clearly a good investment.

Microstructured, drag-reducing coatings for large components (see Chapter 3.4 and Figure 4.2) are an example of a technology of average industrial maturity. The fundamental R&D work is complete, but there is still a major need for application-oriented development work, in particular relating to application techniques, before the technology can be utilised by industry. The fundamental R&D work has already demonstrated the potential and uses of the technology. Issues relating to durability and resistance have been answered or are being investigated. With this level of industrial maturity, about 3 to 5 years is still required for commercialisation of the technology.

Some of the technologies that have been mentioned in this book are at a very early stage of development. The physical or chemical phenomena that are responsible for the specific functions are still in the initial R&D stage. Lab-scale demonstrators are often already available,

Volkmar Stenzel, Nadine Rehfeld: Functional Coatings

ISBN: 978-3-86630-856-5

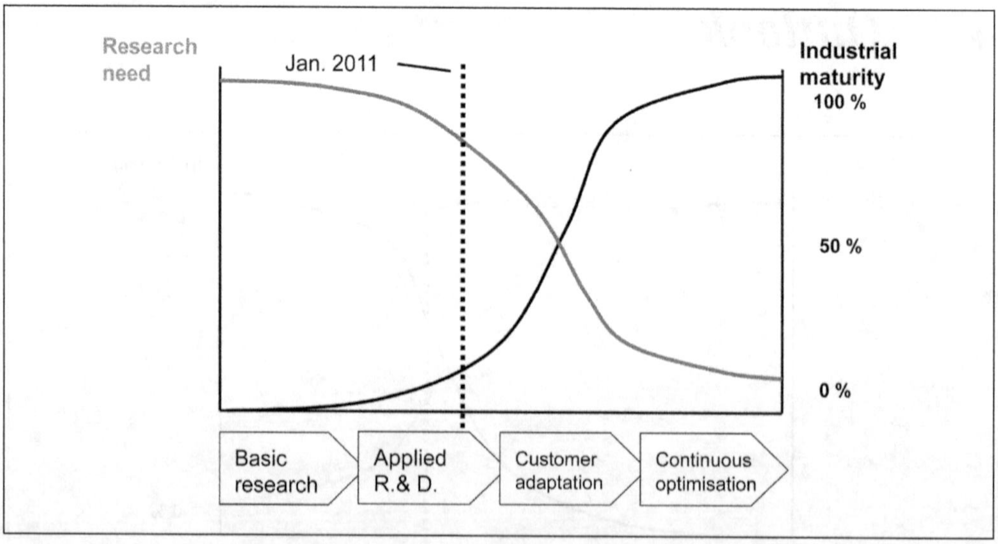

Figure 4.2: Example of a functional coating at a medium stage of maturity: microstructured drag-reducing coatings

but considerable R&D work is still required to determine the potential of the technology for industrial use. The risks associated with such development work are naturally correspondingly high. The time frame here for industrial implementation is 10 to15 years. Biologically modified coatings are one example of a technology at an early stage of development (see Chapter 2.6 and Figure 4.3).

Regardless of the stage of development of a technology, one thing is certain: In the coming years industrial coatings will become ever more important and will make a considerable contribution to adding value to industrial products. In the past coatings were primarily thought of as being for decoration and protection, but in the future the way we think of coatings will be radically different.

Our hope is that this book contributes a little towards this.

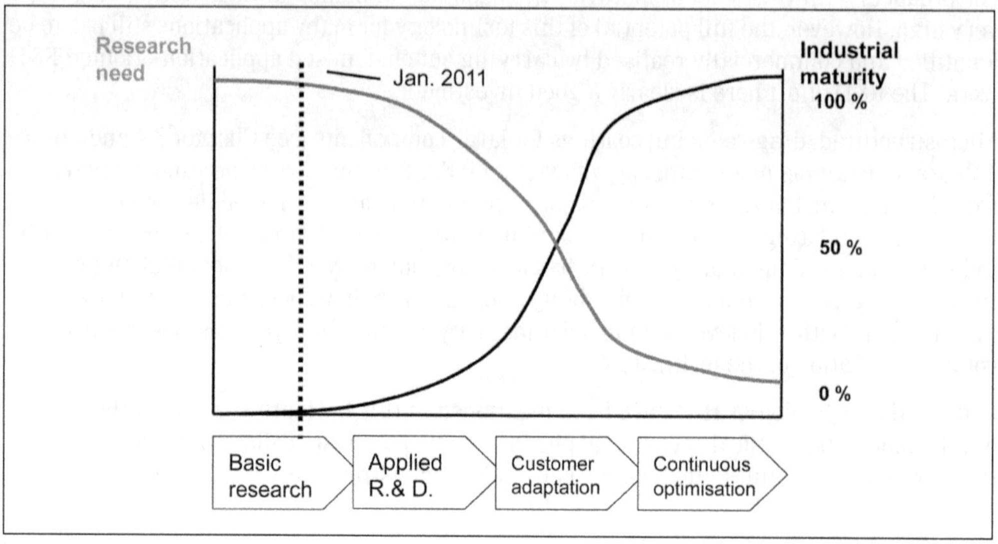

Figure 4.3: Example of a functional coating technology at a very early stage of maturity: biologically modified coatings

Abbreviations

AFP	antifreeze protein
ALD	atomic layer deposition
APS	alkylpyridinium salts
cd	drag coefficient
CAP	cationic antimicrobial peptide
CNT	carbon nanotube
CPP	chlorinated polypropylene
DBTL	di-n-butyltindilaurate
DCPD	dicyclopentadiene
DDSC	dynamic differential scanning calorimetry
DETA	diethylenetriamine
DOPA	dihydroxyphenylalanine
EVA	ethylene vinyl alcohol
FRP	free radical polymerisation
GA	glycoamylase
HDPE	high density polyethylene
HMDA	hexamethylene-1,6-diamine
HMDI	hexamethylene-1,6-diisocanate
HOX	hexose oxidase
IPDI	isophorone diisocyanate
IR	infrared
ITO	indium-tin-oxide
LbL	layer-by-layer
LDH	layered double hydroxide
LDPE	low density polyethylene
LPD	liquid phase deposition
MA	maleic anhydrite
MMT	montmorillonite

Volkmar Stenzel, Nadine Rehfeld: Functional Coatings

ISBN: 978-3-86630-856-5

MWNT	multiwall-nanotube
PAA	polyacrylic acid
PC	polycarbonate
PCM	phase-change material
PDMS	polydimethylsiloxane
PE	polyethylene
PEG	polyethylene glycol
PEI	polyethylene imine
PET	polyethyleneterephtalate
PI	polyimide
PLD	pulsed laser deposition
PMF	poly(melamine-formaldhyde)
PMMA	polymethylmethacrylate
POSS	polyhedral oligomeric silsesquioxane
PS	polystyrene
PTFE	polytetrafluorethylene
PUF	poly(urea-formaldehyde)
PUR	polyurethane
PVA	polyvinylalcohol
PVC	polyvinyl chloride
QLL	quasi liquid layer
Re	Reynolds number
SAM	self-assembled monolayer
SEM	scanning electron microscopy
SPC	self-polishing copolymer
SPPS	solid phase peptide synthesis
SWNT	singlewall-nanotube
Tg	glass transition temperature
TBT	tributyl tin
TDI	toluene-2,4-diisocyanate
TEM	transmission electron microscopy
TEOS	tetraethoxysilane
UV	ultraviolett
VSCC	volatile silicon containing compounds

Authors

Dr. Volkmar Stenzel studied chemistry and started working in the field of paints and coatings in January 1995. After spending several years in different positions in the paint and raw material industries, he joined the Fraunhofer IFAM in Bremen (Germany) in 2001 as head of Paint/Lacquer Technology business unit.

Nadine Rehfeld works in the Paint Technology group at the Fraunhofer IFAM in Bremen, and specialises in multifunctional coatings. She is a biology graduate and started working on anti-microbial coatings in 2006. Anti-icing systems are currently her main focus.

Contact details:

Volkmar.Stenzel@ifam.fraunhofer.de; Phone: +49 (0) 421 / 2246 - 407

Nadine.Rehfeld@ifam.fraunhofer.de; Phone: +49 (0) 421 / 2246 - 432

Volkmar Stenzel, Nadine Rehfeld: Functional Coatings

ISBN: 978-3-86630-856-5

Acknowledgements

We would like to thank everybody who has assisted with the preparation of this book.

Special thanks are extended to the following persons for technical assistance:

Wolfram Hage (DLR, Berlin) for the revision and completion of the "Drag Reducing Coatings" section
Daniel Arnaldo del Cerro (University of Twente) for the information about the creation of microstructures and nanostructures using laser methods
Sabine Scharf (IFAM, Bremen) for the information about micro-encapsulation and for revision of the "Microcapsules" section
Alexej Kreider (IFAM, Bremen) for the information about the coupling of anti-freeze proteins to organic coatings (Chapter 2.6)
Kuna Lukas Steffan and **Andreas Stake** (IFAM, Bremen) for the detailed information on sol-gel systems

We are grateful to the following for making photographs and illustrations available and for providing technical information (names in alphabetical order):

Colin Anderson (International Coatings), **Kellar Autumn** (Portland, Oregon USA), **Wilhelm Barthlott** (Nees-Institute for Biodiversity of Plants), **Werner Baumgartner** (RWTH-Aachen, Germany), **Anne-Grete Becker** (IFAM, Bremen), **Valeria Bers** (Alfred-Wegener-Institute for Polar and Marine Research, Bremerhaven), **Bharat Bhushan** (Ohio State University, USA), **Andreas Brinkmann** (IFAM, Bremen), **Wayne Daniell** (NanoScape AG, Germany), **Rocky de Nys** (James Cook University, Townsville, Australia), **Marko Gasparič** (GRM Ljubljana, Slovenia), **Ingo Grunwald** (IFAM, Bremen), **Wolfgang Luther** (VDI), **Andreas Hartwig** (IFAM, Bremen), **Michael Hofmann** (IFAM, Bremen), **Jörg Ihde** (IFAM, Bremen), **Søren Kiil** (Technical University of Denmark), **Daniel Kolacyak** (IFAM, Bremen) **Ulrike Mock** (Robert Bosch GmbH), **Helmuth Möhwald** (MPI für Kolloid- und Grenzflächenforschung), **Stefan Møller Olsen** (Hempel A/S, Denmark), **Michael Noeske** (IFAM, Bremen), **Thomas Neubert** (IST, Braunschweig), **Rolf Nothhelfer-Richter** (IPA, Stuttgart) **Jörg Rembielewski** (IFAM, Bremen), **Klaus Rischka** (Fraunhofer IFAM), **Dirk Salz** (IFAM, Bremen), **G. Schmidt** (Center for Tropical Marine Ecology, Bremen, Germany), **Stephan Sell** (IFAM, Bremen), **D. G. Stavenga** (University of Groningen, The Netherlands), **Iris Trick** (IGB, Stuttgart), **Petra Uhlmann** (Leibniz Institute of Polymer Research, Dresden)

We would also like to thank **Maribel Stenzel** (IFAM, Bremen) for carrying out the literature survey and **Stuart Fegan** for translating the text.

The book includes results from some of our own research projects which were gratefully funded by the following bodies:

German Federation of Industrial Research Associations e.V. (AiF), Germany

Volkmar Stenzel, Nadine Rehfeld: Functional Coatings

ISBN: 978-3-86630-856-5

Federal Ministry of Economics and Technology (BMWi), Germany

Federal Ministry of Education and Research (BMBF), Germany

The European Regional Development Fond and the Federal State of Bremen, Germany

Volkswagen Foundation, Germany

Our thanks naturally also go to the **Fraunhofer Institute for Manufacturing Technology and Applied Materials Research IFAM,** Bremen, and the **Fraunhofer Gesellschaft e.V.,** Munich.

Last but not least we would like to thank our families. Without their support and patience this book would not have been possible.

Volkmar Stenzel & Nadine Rehfeld

Index

A

B

C

D

Volkmar Stenzel, Nadine Rehfeld: Functional Coatings

ISBN: 978-3-86630-856-5

E

F

G

H

I

L

M

N

O

P

R

S

T

U

W

Y

Z